Early Computer Science Education –
Goals and Success Criteria for Pre-Primary and Primary Education

"Haus der kleinen Forscher" Foundation:

SPONSORED BY THE

Federal Ministry
of Education
and Research

PARTNERS

**Siemens Stiftung**
**Dietmar Hopp Stiftung**
**Dieter Schwarz Stiftung**
**Friede Springer Stiftung**

Scientific Studies on the Work of the "Haus der kleinen Forscher" Foundation

*Volume 9*

"Haus der kleinen Forscher" Foundation (Ed.)

# Early Computer Science Education – Goals and Success Criteria for Pre-Primary and Primary Education

Nadine Bergner, Hilde Köster, Johannes Magenheim,
Kathrin Müller, Ralf Romeike, Ulrik Schroeder, Carsten Schulte

With a foreword by Ilan Chabay

Verlag Barbara Budrich
Opladen • Berlin • Toronto 2023

Edited by: "Haus der kleinen Forscher" Foundation
Responsible editor: Dr Janna Pahnke
Project lead: Dr Elena Harwardt-Heinecke
Conception and editing: Dr Claudia Schiefer
Editorial assistance: Lisa Gerloff, Nina Henke, Katrin Volkmann
Translation: Proverb oHG

Further Information can be found at: https://www.haus-der-kleinen-forscher.de/en/

> Do you have any remarks or suggestions regarding this volume or the scientific monitoring of the Foundation's work?
> Please contact: forschung@haus-der-kleinen-forscher.de.
> Further information and study findings can also be found at https://www.haus-der-kleinen-forscher.de/en/, under the heading "Research and Monitoring".

© 2023 This work is licensed under the Attribution-NonCommercial-NoDerivatives 4.0 International (CC BY-NC-ND 4.0) License. To view a copy of this license, visit https://creativecommons.org/licenses/by-nc-nd/4.0/

You are free to share, copy and redistribute the material in any medium or format. Commercial use and modification only with permission by Verlag Barbara Budrich. Excluded from this license are all illustrations and photographs, whose use outside the narrow limits of copyright law is inadmissible without the consent of the publisher and is punishable by law. This applies particularly to copies, translations, microfilming and the storage and processing in electronic systems.

© 2023 Dieses Werk ist bei Verlag Barbara Budrich erschienen und steht unter folgender Creative Commons Lizenz:
https://creativecommons.org/licenses/by-nc-nd/4.0/deed.de

This book is available as a free download from www.budrich.eu (https://doi.org/10.3224/84742646).

|      |                                |
|------|--------------------------------|
| ISBN | 978-3-8474-2646-2 (Paperback)  |
| eISBN| 978-3-8474-1816-0 (PDF)        |
| DOI  | 10.3224/84742646               |

Verlag Barbara Budrich GmbH
Stauffenbergstr. 7. D-51379 Leverkusen Opladen, Germany
86 Delma Drive. Toronto, ON M8W 4P6 Canada
www.budrich.eu

A CIP catalogue record for this book is available from
Die Deutsche Bibliothek (The German Library) (http://dnb.d-nb.de)

Jacket illustration: Bettina Lehfeldt, Kleinmachnow, Germany
Cover Picture Credits: Christoph Wehrer/"Haus der kleinen Forscher" Foundation
Language editing: Marc Weingart, Werther – http://www.english-check.de
Typeset by Ulrike Weingärtner, Gründau, info@textakzente.de
Printed in Europe on acid-free paper by Elanders GmbH, Waiblingen

# Contents

About the Authors ................................................. 9

Preface .............................................................. 11

Foreword ........................................................... 13
*Ilan Chabay*

Introduction ....................................................... 19
*"Haus der kleinen Forscher" Foundation*

1    Overview of the "Haus der kleinen Forscher" Foundation ........... 20
2    Relevance of Early Computer Science Education ................... 32
3    Professional Basis for the Subject Area of "Computer Science" ...... 34

Summary of Key Findings ........................................ 37
*"Haus der kleinen Forscher" Foundation*

A    **Goal Dimensions of Computer Science Education at the Elementary and Primary Level** .................................. 41
*Nadine Bergner, Hilde Köster, Johannes Magenheim, Kathrin Müller, Ralf Romeike, Ulrik Schroeder, Carsten Schulte*

1    **Potential of Computer Science Education** ......................... 42
1.1   What is Computer Science? ...................................... 43
1.2   Computer Science as a Science .................................. 43
1.3   Construction in Computer Science ............................... 44
1.4   Similarities and Differences in Computer Science in Comparison ... 53
1.5   Computer Science and Computer Science Education .............. 58
1.6   The Relationship of Computer Science Education, Media Education & Digital Education ............................ 70
1.7   Conclusion: Computer Science Education for all .................. 74

2    **Foundation of Goals on the Children's Level** ....................... 76
2.1   Children in Digital Worlds ........................................ 76
2.2   Foundations of Learning Psychology ............................. 82
2.3   Access to Computer Science for Children ......................... 84
2.4   International Comparison: Curricula and their Classification in the Competence Model ............................................... 98
2.5   Placing the International Standards Within the Framework of a Competence Model for Computer Science Education at the Primary Level ............................................. 116
2.6   Results/Conclusion .............................................. 128

| 3 | Goals at the Level of the Children | 130 |
|---|---|---|
| 3.1 | Overarching Basic Competencies | 131 |
| 3.2 | Motivation, Interest and Self-Efficacy of Computer Science | 132 |
| 3.3 | Computer Science Competencies of Children | 135 |
| 3.4 | Prioritisation of Specific Competence Expectations at the Level of the Children | 146 |
| 4 | **Goals for Early Childhood Educators and Primary School Teachers** | 158 |
| 4.1 | Motivation, Interest and Self-Efficacy | 161 |
| 4.2 | Attitudes, Approaches and Understanding of Roles | 162 |
| 4.3 | Computer Science Competencies | 165 |
| 4.4 | Computer Science Didactic Competencies | 182 |
| 4.5 | Key Competencies for Dealing With Digital Media | 196 |
| 4.6 | Conclusion/Recommendations | 198 |
| 5 | **Examples of Prioritised Competence Domains for Computer Science Education** | 199 |
| 5.1 | Examples of Early Computer Science Education | 200 |
| 5.2 | Summary Heat Map of Priority Setting in the Examples | 224 |
| 6 | **Prerequisites for Successful Early Computer Science Education** | 226 |
| 6.1 | General Conditions for Successful Implementation | 226 |
| 6.2 | Measuring Instruments to Determine the Prerequisites for Success | 230 |
| 7 | **Conclusion** | 234 |

## B  Professional Recommendations for Informatics Systems ... 237
*Nadine Bergner, Kathrin Müller*

| 1 | Introduction | 238 |
|---|---|---|
| 2 | Overview of Possible Informatics Systems | 239 |
| 3 | Description and Technical Assessment of Individual Informatics Systems | 241 |
| 3.1 | Cubetto Robot from Primo Toys | 241 |
| 3.2 | Beebot from Terrapin | 243 |
| 3.3 | KIBO from KinderLab Robotics | 245 |
| 3.4 | Ozobot/Ozobit from Evollve Inc. | 248 |
| 3.5 | LEGO WeDo 2.0 | 251 |
| 3.6 | Dash & Dot from Wonder Workshop | 253 |
| 3.7 | Scratch and ScratchJR | 255 |
| 3.8 | Makey Makey from JoyLabzLLC | 257 |
| 3.9 | LEGO Mindstorms (NXT & EV3) | 259 |
| 3.10 | Arduino Microcontroller with ArduBlock | 261 |

| 4 | Recommendations | 264 |
| 5 | Conclusion | 267 |

**Conclusion and Outlook – How the "Haus der kleinen Forscher" Foundation uses the findings** ............................................. 269
*"Haus der kleinen Forscher" Foundation*

| 1 | Recommendations from the Expert Reports as a Basis for the (further) Development of the Foundation's Substantive Offerings | 270 |
|---|---|---|
| 1.1 | Motivation, Interest and Self-Efficacy when Dealing with Computer Science | 271 |
| 1.2 | Computer Science Process Domains | 275 |
| 1.3 | Computer Science Content Domains | 279 |
| 1.4 | Computer Science Didactic Competencies | 282 |
| 1.5 | Attitudes, Mindsets and Understanding of Roles with Regard to the Design of Computer Science Education | 285 |
| 2 | Digital Education – A Chance for Good Early STEM Education for Sustainable Development | 287 |
| 3 | Scientific Monitoring and Evaluation of the Professional Development Workshops | 290 |
| 4 | Outlook – Organisational Development in Educational Institutions | 293 |

**References** ................................................. 297

**Appendix** .................................................. 319

**Illustration Credits** ......................................... 334

**"Haus der kleinen Forscher" Foundation** .......................... 335

**English Publications issued by the "Haus der kleinen Forscher" Foundation to date** ............................................ 337

## About the Authors

### Prof. Dr Nadine Bergner
Technische Universität Dresden, Faculty of Computer Science
*Research interests:* Subject didactics of computer science, out-of-school learning, school laboratory, teacher training and advanced education, digital education, computer science in primary schools, learning games, learning technologies
*Contact:* nadine.bergner@tu-dresden.de

### Prof. Dr Ilan Chabay
Research Institute for Sustainability Helmholtz Centre Potsdam
*Research interests:* Processes of societal change toward just and equitable sustainable futures, understanding functions of narrative expressions of vision and identity in collective behavior change, using narratives in modeling social dynamics, collaborative creativity and its role in innovation for societal needs, design of games that stimulate curiosity in global change and sustainability
*Contact:* ilan.chabay@gmail.com

### Prof. Dr Hilde Köster
Freie Universität Berlin, Department of Education and Psychology
*Research interests:* Primary school pedagogy, social studies and science in primary education, educational processes and diagnosis-based giftedness support for children in the fields of natural sciences, technology and computer science, professionalisation of prospective primary school teachers.
*Contact:* hkoester@zedat.fu-berlin.de

### Prof. Dr Johannes Magenheim
Paderborn University, Faculty of Electrical Engineering, Computer Science and Mathematics
*Research interests:* Computer science didactics, computer science and education, e-learning
*Contact:* jsm@uni-paderborn.de

### Kathrin Müller
ATIW Berufskolleg Paderborn
*Research interests:* Computer science education, learners' understanding of robots and how they work, teacher training, in-service teacher training in computer science at primary level
*Contact:* kathrin.mueller@atiw.de

### Prof. Dr Ralf Romeike

Freie Universität Berlin, Institute of Computer Science
*Research interests:* Computer science in early childhood education, agile methods in project-based teaching, visual programming languages, data literacy, artificial intelligence education
*Contact:* ralf.romeike@fu-berlin.de

### Prof. Dr Ulrik Schroeder

RWTH Aachen University, Department of Computer Science – Learning Technologies & Computing Education
*Research interests:* Theories, methods and tools for learning technologies, learning analytics, assessment and feedback, game-based learning, mobile learning, subject didactics of computer science, extracurricular learning venues, development and research of learning materials for computer science education and for extracurricular learning venues, computer science in primary schools
*Contact:* ulrik.schroeder@rwth-aachen.de

### Prof. Dr Carsten Schulte

Paderborn University, Faculty of Electrical Engineering, Computer Science and Mathematics
*Research interests:* Computer science didactics, concepts of computer science education for the digital world, acquisition, understanding and use of digital devices and infrastructures, primary school learners' understanding of informatics systems and how these can be changed through exposure to the algorithmic functioning of the devices
*Contact:* carsten.schulte@uni-paderborn.de

# Preface

**Dear Readers,**

Whether traffic lights, navigation devices in cars, parents' smartphones, tablets at the child-care centre or PCs in the classroom – children today grow up in a world that is significantly shaped by digital technology and is developing rapidly. Girls and boys want to explore and help shape it and have many questions: How does a robot work? What happens when I switch on the computer? And where do all the pictures and information come from?

The "Haus der kleinen Forscher" Foundation has been successfully developing and evaluating concepts and materials in the STEM field for exploration- and inquiry-based learning in early childhood education for many years. In 2018, the Foundation completed its groundwork in the areas of S-T-E-M with a volume on early computer science education, thereby laying the basis for its work on the topic of Computer Science. I am pleased that this English translation will allow us to provide our international readership with a slightly updated version of the volume.

By establishing the subject-specific basis for early computer science education, we have ventured into new territory, since there are still hardly any research approaches or educational concepts on this topic in the German-speaking world. At the same time, the importance of computer science and the competencies associated with it is steadily growing – something that became even clearer in the Coronavirus pandemic.

As far as children are concerned, the prevailing urge is a fascination with the world around them: they want to try things out, explore, participate and find out how things work and how they are interconnected. In doing so, they approach computers and other devices no differently than they do other intriguing things. How does all this information get into a little mobile phone? Why is my computer so fast? How do the traffic lights know when to indicate that cars or pedestrians are to stop or move?

Our task as an early education initiative is to enable educators to search for answers with the children.

For this reason, the "Haus der kleinen Forscher" Foundation focuses on the technological perspective of digital aspects in its educational offerings. There are now well-developed professional development concepts, educational materials and online offerings available for early childhood educators, primary school teachers and children that enable them to discover computer science – with and without computers.

The aim is not to increase the use of digital media but to understand the underlying concepts. Children can learn about algorithmic thinking in informatics systems hands on, thereby laying an important basis for reflective and competent use of these systems.

Pioneering work has been done in the field of early computer science education – I would like to express my sincere thanks to the authors of this volume for their great support and guidance in this exciting terrain. A responsible, open-minded and creative approach to computer science is important for our children and for the society of tomorrow.

I strongly believe that this volume and our educational offerings are able to contribute to ensuring that early computer science education is no longer uncharted territory in educational institutions throughout Germany, but a valuable part of good and successful early STEM Education for Sustainable Development.

Michael Fritz
Chairperson of the Executive Board of the "Haus der kleinen Forscher" Foundation

# Foreword

*Ilan Chabay*

The first stanza of Bob Dylan's 1964 song ends with

"And you better start swimmin'
Or you'll sink like a stone
For the times they are a-changin'"

It has always been true that the times are changing, but what we humans face now are changes that challenge us in new and profound ways - ways for which many previously successful systems and approaches are no longer adequate. Thus, we need to continually find and learn new ways to swim in the fast-moving, turbulent stream we all inhabit on our shared planet Earth.

Finding and following culturally and contextually adaptive pathways to sustainable futures, including addressing climate change, is indeed a profound challenge. Successfully confronting a challenge of this magnitude and scope (e.g., as aspired to in the United Nations Sustainable Development Goals for 2030) requires not only that we use the best science and technology available. We must also learn to engage meaningfully in the collective decisions and actions to avoid, mitigate, or adapt to changing conditions in ways that are attuned to our diverse cultures and contexts.

Computer science and technology have already helped us immensely to understand the changes occurring on Earth with its limited resources and to slowly recognise the urgent need to change unsustainable patterns in society. At the same time, digital systems, vast data bases, and applications of rapidly increasing computational power (e.g., cell phones, navigation systems, and computers) have become nearly ubiquitous in society. But we have not yet equipped most young people in our societies to understand computer science and technologies, to use them most effectively for their purposes, and to play a role in the design of such systems and decisions whether to develop and deploy them or not. Nor have we made use of the powerful intellectual building blocks at the core of computer science to improve thinking strategies and skills that can be applied effectively not only in STEM (science, technology, engineering, and math) but in many aspects of our lives.

Why not? To incorporate computer science in STEM learning, we must overcome the widespread perception that computer science is a narrow domain of specialisation for "nerds" and "geeks" – neither relevant nor accessible to most adults and especially not to young children. This obstacle exists despite the nearly ubiquitous use, even by young children, of digital devices and computation in our daily lives. More to counter this point is in the work described in this volume of the "Haus der kleinen Forscher" Foundation, showing that young children are able to learn from aspects of computer science in developmentally appropriate ways. The misperception and failure to introduce the ideas and experiences is an obstacle to anyone, but it more frequently adversely affects girls and underserved and marginalized populations.

On the other hand, why introduce this domain of learning? Is it essential that these obstacles be overcome? What drives the "Haus der kleinen Forscher" Foundation and others to make the considerable effort to research the field and develop strategies and materials to introduce computer science to STEM learning for young children in and out of school?

There are several answers to these questions. At a meta level, offering developmentally appropriate experiences with computer science and digital systems engages children's curiosity and rewards it with positive, playful experiences. Such experiences and explorations build confidence and desire for learning about processes and content that affects the children's lives now and will do so in new ways in their future. This may form a strong platform not only for understanding and using digital media, devices, and computational power, but the interest and confidence to develop more fully their capacity to design and shape that future for their benefit and that of their society.

Learning with and about computer science and its manifestations from an early age through curiosity and play provides an excellent natural entrée to acquiring thinking skills of great value in every area of STEM (and indeed in all aspects of life!). The point is not only to learn about specific systems, processes, or devices, but to develop powerful learning strategies in an increasingly sophisticated way in order to approach new and more complex problems. This process can be started very simply by setting an objective and learning to dismantle the task into manageable steps leading to the objective. How do we get the penguin in the corner square on the grid to the delicious dinner in the centre and not bump into the seal or ice block on the way? We can try sequences of directional steps to reach the objective. This and other exercises can be done with an electronic robot, with paper and pencil on a grid, or children moving on a grid taped to the floor.

Even more complicated systems can be playfully explored in an elementary school classroom. For example, build a working model of the internet – no computer or electronic devices needed. Messages, each one sentence long, are written on strips of paper. Then the messages are cut up into information packets that are sent across the room to their chosen destination. The packets travel by courier (children) from one way-station (where a child acts as router) to another to the final reception station where the packets are reassembled into a semblance of the original messages. The fun lies in the activity that engages the children (and teachers) in the formation of a model system with children as routers and carriers, designing and executing stepwise iterative instructions for the carriers of the packets of information, and laughing at and eventually troubleshooting the inevitably garbled messages that appear when the packets of pieces are reassembled into messages. This activity, which I designed for the US National Engineering Week in 2001 for the National Academy of Engineering and later used in elementary classrooms, is illustrative of the point made in this volume (p. 171):

> "[B]asic principles of communication on the Internet should be explored 'unplugged' using simple playful means, since a basic understanding of how the Internet works is an important part of computer science education. Suitable examples include role-plays on sending mail via routers,…".

These examples are very simple, but even so, they illustrate a fundamental facet of human thinking and learning, namely the use of models. Models may be tacit and procedural, as in learning from experience where to place your hands to catch the ball, or they may be highly complex computational models of natural or socio-ecological systems that inform our understanding of our challenges of achieving more sustainable futures. It is important to recognise that models are an essential part of how all humans process sensory data that we constantly receive. We need to make sense of inputs to make decisions in different situations that could be consequential – a glimpse of the on-coming truck as one start to cross the street – or on complex systems like the weather office estimating the likelihood of a flash flood occurring in a particular area. The critical issue is that our brains have a working memory bank that is unable to retain and process more than a few variables at any one time. Complex systems may have scores or even thousands of interacting components and variables in play. To begin to unpack and examine the options and consequences of potential decisions, models are essential tools. It is through decades of work by many scientists around the world using increasingly powerful computer models that we are making progress in many fields of science, as well as opening up new directions in music, visual arts, improved accessibility for the visual or auditorily impaired, and enriching or verifying historical records.

There are many types of analytical and computational models in use, but in most education settings from the primary through tertiary level, students are shown only the results of models rather than explicitly discussing the models that lead to the results. Learning computer science in early education, including the explicit processes of making and using different kinds of models, will help students develop and use models in their studies and better understand the influence of models in their livelihoods and private lives as citizens. Understanding the role of models and their value figures prominently not only in STEM but also for interpreting complex evidence needed for informed deliberation for decision-making on policy and practice in society. Learning the processes and content of computer science opens students to a wider range of possible livelihood choices as they develop. It also increases their understanding of the roles and impact that digital systems have in daily life.

Models are also the engines of the games that many children play on phones, tablets, and computers. As such, the games can be an appealing avenue into aspects of computer science. In a constructivist approach, children's interest in games provides a stimulus for deconstructing the operations and user interface of a game and then modifying the game or creating one's own game. The simple set of sequential instructions for moving a penguin around a space can become the starting point for a very elementary, "unplugged" game played on a physical two-dimensional grid, with one person "programming" the penguin's moves in coordinated time steps after the opponent decides where to place the hungry seal in order to try to intercept the penguin. That physical version can be developed into more complex and dynamic forms and be coded into a simple app as the students become more adept. This algorithmic process is at the core of animations and programming that students will encounter.

Depending on each child's interest and level, more sophisticated games with immersive environments, complicated logical branching structures, narrative flow, and tailored user interfaces become accessible for analysing, modifying, and emulating. If the basis for awareness of the computer modeling and processes operating beneath the surface of the game is introduced in early learning, electronic games become more valuable as environments for learning and creativity. As the children develop as learners, the activities can become more sophisticated and the connections to algorithms, coding, apps, models and games further developed.

The authors of the first expert report of this volume comment on the concern that media and games often are decoupled from learning by stating:

> "In our opinion, a special role of schools and possibly also childcare centres, could be to promote and point out further possibilities and types of interaction in addition to the predominant consumption activities. In addition, interactions could not only be experienced in isolation and individually, they could be experienced by parents and children together, so that the experiences can be verbalised and processed, as well as reflected on and classified in an age-appropriate way" (Bergner et al., p. 128).

The point made in the above quote about the experience as a group activity suggests circling back to emphasise the earlier concern about resistance or obstacles to introducing computer science in early learning venues. Engaging in designing or playing games that explicitly build on computer science processes are activities that in many cases can be done as group collaborations rather than only as individual efforts. This may help students, teachers, and parents see learning computer science less stereotypically as an asocial, strange, alienating activity and more as a socially connected one. Motivating collaborative learning opportunities in computer science can be done with many interesting project challenges at developmentally appropriate levels for young children. Developing the skills and pleasure for collaborative work from an early age, whether through computer science or other learning activities, is tremendously important. It is indispensable for finding interdisciplinary solutions for the many complex challenges we face.

In addition to finding time in the elementary classroom schedule and re-assigning priorities to include computer science, teachers who are adept and comfortable with the elementary content and processes of computer science are obviously needed. But a classroom with adequate materials and support is also a learning space for teachers willing to be co-learners with their students and colleagues. A wider introduction of computer science in early learning will require pre-service and in-service education of teachers and expert mentoring, but it will also be augmented by classroom experience. As parents and teachers, many of us have already learned from children who have already so rapidly acquired digital skills.

As stated on page 75 of this volume, "they should not only learn to deal specifically with a digital artefact, but also to confidently master general and transferable strategies for exploring an unknown system, while also thinking about its possibilities, limits and effects". It is not only a matter of understanding and thus being able to make better use of these artefacts and devices as children develop, but of becoming adept and comfortable in using what is learned in one domain and applying or adapting it to a problem in another domain.

Gaining domain knowledge and skills, as well as confidence in one's capacities learned in early childhood experience with computer science in STEM has great value beyond STEM. The same knowledge and sense of confidence is essential for making informed normative choices of safe and ethical use of technology but also for health and behavioural choices, which children will face even before becoming adults.

The thoughtful and evidence-based approach to computer science learning in STEM education of young children that is presented in this volume may have long term positive effects on the development of those children throughout their lives. One particular hope it raises for me is that the thinking skills, reliance on evidence-informed decision-making, and sense of personal agency with computer science and digital systems will provide a bulwark against the grip of baseless conspiracy theories and misinformation that threatens our societies.

The ninth volume in the series from the "Haus der kleinen Forscher" Foundation lays out a thorough and important examination of the rationale, educational framework, and materials for introducing computer science into the learning experiences of young children from pre-school through elementary school age. The needs of teachers of young children in and outside of school are addressed, as are the reasons that it remains difficult to engage some children, parents, and teachers in supporting the expanded offerings. The volume is not only a rich source of information and ideas for implementing computer science learning in STEM. It is also an inspiring call to action for parents and teachers alike to support the leap toward a future that helps all people to swim well in the turbulent stream that is our rapidly changing social-ecological system on Earth.

Prof. Dr Ilan Chabay
Research Institute for Sustainability Helmholtz Centre Potsdam

# Introduction

"Haus der kleinen Forscher" Foundation

1  Overview of the "Haus der kleinen Forscher" Foundation
2  Relevance of Early Computer Science Education
3  Professional Basis for the Subject Area of "Computer Science"

# 1 Overview of the "Haus der kleinen Forscher" Foundation

Since 2006, the non-profit "Haus der kleinen Forscher" (Little Scientists' House) Foundation has been committed to improving education for children between the ages of three and ten in the domains of science, technology, engineering/computer science, and mathematics (STEM). Together with its local network partners, the Foundation offers a continuing professional development programme throughout Germany that supports early childhood educators and primary school teachers in nurturing children's spirit of discovery and in facilitating their exploration and inquiry activities in a qualified way. The education initiative thus makes an important contribution to improving educational opportunities, fostering the next generation of professionals in the STEM fields, and professionalising pedagogical staff.

The educational initiative thus makes an important contribution in the following areas:

- qualification of early childhood educators
- quality development of institutions
- development of the children's personalities, abilities and interests
- promotion of the next generation of professionals in STEM educational fields

The main activities of the Foundation are:

- establishment and expansion of sustainable local networks with the participation of local stakeholders as well as counselling and support of the now more than 200 network partners,
- training multipliers (local trainers) who provide ongoing guidance for local early childhood educators, teachers and leaders,
- development and provision of professional development concepts and materials for early childhood educators, teachers and leaders,
- supporting the quality development of educational institutions based on "Haus der kleinen Forscher" certification, as well as
- the evaluation and scientific monitoring of the Foundation's activities.

## Qualification initiative for educators

The "Haus der kleinen Forscher" Foundation is Germany's largest early childhood education initiative in the domains of science, technology, engineering/computer science, and mathematics. It supports child-care centres, after-school care centres and primary schools in setting mathematical, computer science, scientific and/or technical priorities and establishing education for sustainable development (ESD), as well as creating a conducive development and learning environment for children. The Foundation's educational approach builds on the children's resources and emphasises joint exploration and inquiry in dialogue-based exchange (Stiftung Haus der kleinen Forscher, 2019a). Through its activities, the Foundation also promotes the implementation of the existing educational and framework curricula of the respective federal states in the areas of science, technology, engineering/computer science, mathematics and supports the anchoring of ESD in the educational areas.

The Foundation's content-related offers include **professional development** for early childhood educators, teachers and leaders, as well as educational materials, an annual activity day and suggestions for cooperation:

- **Educational materials:** For practical implementation in educational institutions, the Foundation provides printed and online materials (available at: haus-der-kleinen-forscher.de) free of charge in its professional development courses, e.g., topic brochures, exploration and inquiry cards, didactic materials and film examples.
- **Website:** The website haus-der-kleinen-forscher.de provides information for all interested parties about the educational initiative: including the educational programme, contact with local training providers, inquiry ideas for children, scientific studies, brochures and picture cards for exploration and inquiry.
- **Campus:** On the online learning platform campus.haus-der-kleinen-forscher.de, early childhood educators, teachers and leaders can take part in open or moderated online courses and on-site training, or exchange ideas with each other in various forums. All of the Foundation's online services are available free of charge
- **Blog:** The "Haus der kleinen Forscher" blog (blog.haus-der-kleinen-forscher.de) offers a forum for dialogue on good early STEM education for sustainable development. Here, the Foundation presents itself as a player in the national and international world of education and foundations while at the same

time providing a detailed look behind the scenes of the "Haus der kleinen Forscher".

- **Social media channels:** On Facebook, Twitter, Instagram and YouTube, the Foundation shows what good early STEM education can look like: *@kleineForscher* (on Instagram at *kleine_Forscher*). Here, current reports and studies are shared, videos of ideas and activities for exploration and inquiry are shown and information about the Foundation's latest educational offers is provided.

- **"Forscht mit!" magazine:** Every quarter, early childhood educators, teachers and leaders receive practical tips on exploration and inquiry at their institution, information on the Foundation's work and best practice reports from other institutions and networks.

- **"Tag der kleinen Forscher" (Little Scientists' Day):** On this interactive day, children all over Germany can explore current topics of inquiry. For this purpose, the Foundation provides educational institutions with material and calls on supporters from politics, business, science and society to join in.

- **Suggestions for cooperation:** Interested parents, sponsors and other educational partners support joint exploration and inquiry at the institutions.

- **Certification:** Committed institutions are certified as "Haus der kleinen Forscher" based on defined assessment criteria. All institutions that apply receive detailed feedback with suggestions for further developing joint exploration and inquiry with children.

- **Website for children:** At meine-forscherwelt.de, children of primary school age can access an interactive inquiry garden that encourages them to embark on independent journeys of discovery. Tips are available for educators on how to support learning.

- **Service portal integration:** At integration.haus-der-kleinen-forscher.de, early childhood educators, teachers and leaders receive support in the integration of refugee children at child-care centres, after-school care centres and primary schools through a variety of materials, practical ideas and a stimulating exchange of experiences.

## Germany-wide networking

As a Germany-wide educational initiative, the "Haus der kleinen Forscher" thrives on the commitment of a wide range of local stakeholders – the local networks that act as permanent partners and training providers in the regions. There are currently 196 network partners, including municipalities and child-care sponsors, trade associations, science centres, museums, companies, foundations, associations etc. Since 2011, the initiative's professional development programme has also been open to after-school care and all-day primary schools.

Approximately 86,000 early childhood educators and primary school teachers from around 35,000 child-care centres, after-school care centres and primary schools have already participated in the initiative's professional development programme; among these are early childhood educators from around 28,000 child-care centres as well as educators and teachers from around 1,700 after-school care centres and around 5,300 (all-day) primary schools.

Across Germany, around 6,000 child-care centres, after-school care centres and primary schools are certified as "Haus der kleinen Forscher", of which more than 5,400 are child-care centres. Since autumn 2013, after-school care centres and primary schools have also been able to obtain certification. More than 200 after-school care centres and over 300 primary schools have received the "Haus der kleinen Forscher" certificate since then (as of 21 October 2022).

## The continuing professional development programme

The "Haus der kleinen Forscher" Foundation focuses on advanced qualification for educators in discovering and exploring mathematics, computer science, scientific and/or technical topics with children. Since 2018, advanced training courses have been offered with a focus on "education for sustainable development" (ESD). The aim is to provide continuous support for early childhood educators, teachers and leaders: Participation in advanced training courses on various topics successively expands participants' methodological repertoire and deepens the understanding of the Foundation's pedagogical approach. Alternating between face-to-face training and transfer phases, educators can try out what they have learned in practice and exchange ideas at the next training session.

In addition, the Foundation offers a constantly growing range of online courses which anyone interested can use individually, flexibly and free of charge to refresh or deepen their grasp of the subject matter. These include both open online courses, which can be attended independently at any time, and moderated online courses at fixed times. Here, participants work together on the content while the moderator supports them. Web-based seminars (webinars) are scheduled for a specific time and include an interactive online talk. There are also topic-specific forums that allow professionals to reflect on their own practical experience together with other early childhood educators and primary school teachers.

To allow as many interested early childhood educators, teachers and leaders as possible to participate in the professional development programme, further qualification takes place via a multiplication model: The "Haus der kleinen Forscher" Foundation trains trainers at several locations in Germany. These trainers in turn offer professional development for educators in their local network (Stiftung Haus der kleinen Forscher, 2019b). By participating in the Foundation's face-to-face and online training courses, these trainers qualify to offer professional development courses for educators. They receive support in the form of detailed working materials for their task in adult education and also have the opportunity to gain personal feedback through the Foundation's training observation programme or in the form of video coaching. In addition, an online campus is available for the trainers to allow them to refresh and deepen their grasp of the subject matter. The digital learning platform offers a variety of online learning programmes, as well as content-related information and working documents for the individual training modules. For certain topics, there is the possibility to work independently on open e-learning modules, participate in tutor-led courses and use the online support courses for face-to-face professional dvelopment. In addition, trainers can get in touch with each other and exchange ideas in specific forums or open chats. The educational initiative offers different topics for trainers every year.

Since 2017, the Foundation has offered flexible entry to its educational programme for early childhood educators, teachers and leaders, as well as for trainers[1]. If the learning support staff see a need to expand their pedagogical competencies or would like to get an overview of the Foundation's pedagogical concept, they can start with the face-to-face professional development courses "Exploration and Inquiry with Water" or "Exploration and Inquiry with Air", in which the Foundation's pedagogical approach is explored in greater depth; or they can attend the seminar or online course "Basic Seminar – The Pedagogical Approach of the 'Haus der kleinen Forscher' Foundation". Likewise, early childhood educators,

---

1   By making the entry to the Foundation's educational programme more flexible, the Foundation assumes greater personal responsibility for its target groups. In accordance with the notion of the independent learner, which forms the basis of the Foundation's pedagogical concept, it relies on the fact that the early childhood educators, teachers, leaders and trainers can themselves recognise where they stand in terms of their interests and needs and which topic or format is the right entry point for them to the "Haus der kleinen Forscher" educational programme or which offering they wish to use for their further professional development. In order to support users in their professionalisation as learning facilitators in the best possible way, the Foundation offers targeted educational programmes and continuously develops these further in an impact-oriented and needs-based manner.

teachers and leaders or the trainers can choose another module on mathematical, computer science, scientific or technical topics or on education for sustainable development as an introduction. The content is offered in different formats: on-site training, self-education formats (such as online courses or printed educational materials) and educational events. The "Haus der kleinen Forscher" certificate also supports quality development of the educational work at the institutions and makes the commitment to good early STEM education visible to the outside world. The Foundation is strongly oriented towards the needs, prior knowledge, previous experience and interests of its target groups.

The approach of providing support that is as individual and needs-oriented as possible – which the Foundation pursues both at the level of children and early childhood educators, teachers and leaders – is also implemented at the level of the trainers through the quality system for further training. The main elements of the system are the application and accreditation at the beginning of the trainer's activity, qualification phases designed according to needs and re-accreditation every two years. Based on the goal dimensions for multipliers of early STEM education, it systematises and expands the requirements for trainers and makes them explicit (Stiftung Haus der kleinen Forscher, 2019b).

The Foundation's offerings for early childhood educators, teachers and leaders are also developed based on sound goal dimensions relating to subject-specific criteria. They specify which goals are to be achieved in each case. Goal dimensions for children and early childhood educators and primary school teachers have been developed collaboratively with experts for the individual STEM disciplines, as well as for education for sustainable development and used as an orientation basis for the development of courses ("Haus der kleinen Forscher" Foundation, 2018; Stiftung Haus der kleinen Forscher, 2015, 2017a, 2018a, 2019).

The goal dimensions for early computer science education were developed and published in German for the first time in 2018 and are presented in this volume in English translation. At the beginning of the 2017/2018 academic year, the content of the Foundation's offers was expanded – to include the educational area of computer science education – with the workshop "Discovering computer science – with and without computers". In 2021, this was supplemented with the blended learning advanced training course "Computer science education in primary school teaching" (cf. chapter Conclusion and Outlook). All offers were developed based on the present expert report "Goal dimensions of computer science education at elementary and primary level".

In 2018, the Foundation also expanded its range of training courses, content and materials for education for sustainable development which is aimed not only at early childhood educators and primary school teachers, but also, for the first time, at child-care centre leaders. The latest training programme, "Rethinking

Consumption", offers educational staff and leaders at child-care centres the opportunity to learn more about consumption and sustainability at their institutions and address the topic with children. In 2019, the Foundation's offer was supplemented with the professional development course "STEM is everywhere", which for the first time addresses all STEM disciplines together and focuses on methods for learning support that apply to all STEM topics. In 2020, the number of new registrations on the online learning platform multiplied due to the Coronavirus pandemic. The Foundation reacted quickly, and within a very short time was able to offer additional online courses on topics such as "Co-constructive learning support", "Philosophising with children" and "Impulses in inquiry". Since 2021, early childhood educators, teachers and leaders have also been able to participate in online training courses via their local networks.

## Scientific monitoring and quality development

All activities of the educational initiative are continuously monitored and evaluated scientifically. The "Haus der kleinen Forscher" Foundation maintains open exchange with science and professional practice and sees itself as a learning organisation.

A comprehensive range of measures serves to ensure and further develop the quality of the "Haus der kleinen Forscher" (see Figure 1). The Foundation's internal quality management covers all important activities and services with its own evaluation measures and comprehensive monitoring. For this purpose, the Foundation draws on a whole range of data sources (such as event-related surveys of network coordinators, trainers and early childhood educators, teachers and leaders); a combination of cross-sectional and longitudinal data enables a view of the current situation and also of important changes in recent years. In order to be able to react flexibly to the Foundation's need for insights, several surveys are conducted with different target groups at different points in time.

The longitudinal perspective plays an increasingly important role in the Foundation's internal evaluation and monitoring measures, also in order to meet the demand for stronger impact orientation. The Foundation provides important results of these measures in its regularly published monitoring report. For example, the 2016/2017 monitoring report uses an impact chain to describe how the education offered by the initiative contributes to improving early STEM education in Germany (Stiftung Haus der kleinen Forscher, 2017b). In the 2018/2019 monitoring report, the cross-sectional analyses are continued and methodically supplemented by results from the longitudinal study (Stiftung Haus der kleinen Forscher, 2020a). Further analyses of these data show that participation in a "Haus der kleinen Forscher" training course also has a longer-term effect on subject-didactic knowledge, motivation and self-efficacy of the educators. The effects of participation in professional development are also related to the individual characteristics of the educators (e.g., initial level of competence and professional experience) as well as to the organisational framework conditions (e.g., STEM supervision ratio) (Stiftung Haus der kleinen Forscher, 2020a, 2022).

The content-related (further) development of new Foundation courses always takes place in a professionally sound manner and in collaboration with science; new Foundation courses are also developed and tested together and in exchange with working professionals. In cooperation with educators from child-care centres, after-school care centres and primary schools, each new training programme is tested in detail before the training concepts and materials are disseminated in the regional networks. In the process, the participating early childhood educators, teachers and leaders test the feasibility of initial practical ideas and provide feedback on the Foundation's support services. Educational concepts are revised and further developed based on this feedback.

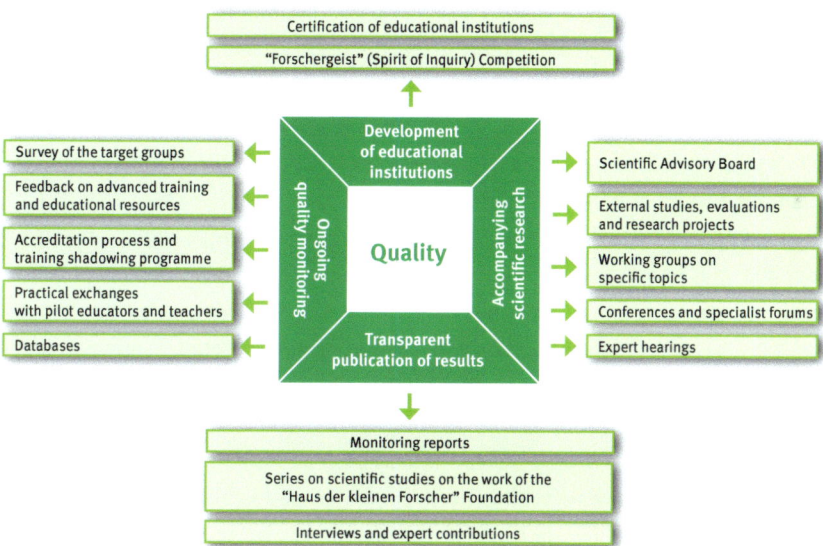

*Figure 1.* Overview of measures to ensure and further develop the quality of the Foundation's services

At the institutional level, "Haus der kleinen Forscher" certification is another important quality development tool (Stiftung Haus der kleinen Forscher, 2020b). The Foundation awards certificates according to a standardised procedure that was developed based on the "Deutsches Kindergarten Gütesiegel" (German kindergarten quality seal) and with the participation of a team of scientists (Yvonne Anders, Christa Preissing, Ursula Rabe-Kleberg, Jörg Ramseger and Wolfgang Tietze).

The reliability and validity of the certification procedure for child-care centres were confirmed in an external scientific study (Anders & Ballaschk, 2014). Certification as a "Haus der kleinen Forscher" is a free procedure for monitoring and enhancing pedagogical quality during the implementation of STEM educational content. By answering the questions in the certification questionnaire and receiving subsequent detailed feedback from the Foundation with practical suggestions and tips, child-care centres, after-school care centres and primary schools are supported in their quality development. Follow-up certification can take place every two years and enables the long-term anchoring and further development of educational quality at institutional level.

From 2012 to 2020, the "Deutsche Telekom" Foundation and the "Haus der kleinen Forscher" Foundation organised the nationwide "Forschergeist" child-care centre competition to honour the commitment of child-care professionals and the quality of institutions and to further motivate them to engage in early education work in the STEM field. In September 2019, the initiators launched the "Forschergeist" competition for the fifth time. Outstanding projects that inspire children for the world of science, technology, engineering/computer science, mathematics as well as education for sustainable development were awarded prizes in 2020. The award-winning projects were documented and published for each of the five competitions so as to serve as good examples to inspire other professionals for exploration and inquiry at child-care centres (Deutsche Telekom Stiftung & Stiftung Haus der kleinen Forscher, 2021)[2].

In addition to continuous monitoring for quality assurance and quality development, the Foundation's work is professionally substantiated and evaluated in research projects as part of long-term external accompanying research conducted with renowned partners. Two independent research groups investigated the impact of science education in early childhood from 2013 to 2017 (Stiftung Haus der kleinen Forscher, 2018b)[3]. The aim of the first research project "Early Steps into Science" (short: EASI Science, funded by the "Haus der kleinen Forscher" Foundation and the Federal Ministry of Education and Research) was to gain insights into the impact of early science education on the scientific competencies of educational staff and children at child-care centres. The results show that early childhood educators who have received further training in science have a higher level of subject knowledge and subject-didactic competence than a comparable group who have not attended such professional development courses. In addition, motivation for and interest in science education are greater among educators

---

2   Documentation of the projects that won the "Forschergeist" competitions of 2012, 2014, 2016, 2018 and 2020 is available at forschergeist-wettbewerb.de.

3   Further information is available at www.haus-der-kleinen-forscher.de under the section "Scientific monitoring".

who have received further training. Children also take more pleasure in learning, show greater interest in science and have more self-confidence in their own abilities if one of the focal points at their child-care centres is science (Steffensky et al., 2018). The second research project "Early Steps into Science and Literacy" (short: EASI Science-L, funded by the "Haus der kleinen Forscher" Foundation, "Baden-Württemberg" Foundation and "Siemens" Foundation) investigated the impact of language education and quality of interaction in the context of science educational programmes. The study showed that inquiry-based learning is well suited to language education. Educators who have attended professional development courses in science create more linguistically stimulating learning opportunities for children than educators without advanced training in this area. The children exhibit a higher level of linguistic competencies if their educators have previously participated in a combined training course on science and language offered by the "Haus der kleinen Forscher" Foundation. In addition, there are positive correlations between the science-related process quality designed by the educator and the children's science skills (Rank et al., 2018).

With a view to the needs-based development of offers, from 2017 to 2019, the Foundation, together with the Federal Ministry of Education and Research, funded a study on the "Entwicklungsverläufe von pädagogischen Fach- und Lehrkräften in der frühen MINT-Bildung" (Early Childhood Educators' and Primary School Teachers' Professional Development; in short: EpFL MINT). The aim of the study was to investigate how self-perceived developments of early childhood educators and primary school teachers can be described in relation to professionalisation in the STEM field and whether different phases or formative events play a special role in the professional development of early childhood educators and primary school teachers. The results show that the number of training courses attended in the "Haus der kleinen Forscher" initiative is of central importance for professional development in early STEM education. The training courses also support educators in breaking down barriers to exploring STEM topics (Skorsetz, Öz, Schmidt & Kucharz, 2020).

Since 2020, the Foundation, together with the Federal Ministry of Education and Research, has been funding a study on the effects of the model programme "KiQ – gemeinsam für Kita-Qualität. Wenn Entdecken und Forschen zum Alltag werden" ("Joining hands for quality in child-care centres. When exploration and inquiry become part of everyday life"). Through this child-care centre programme, the Foundation is looking at the institution as an overall system, adopting a continuing education approach that combines personal and organisational aspects and taking into account that the way an institution is managed and organised has an impact on the pedagogical work at that institution (Deutsches Jugendinstitut & Weiterbildungsinitiative Frühpädagogische Fachkräfte, 2014; Strehmel & Ulber,

2017). During the model phase (2020-2022), the "KiQ" programme will be evaluated by the Foundation and scientifically monitored by an external research group. The aim is to gain important insights into the success indicators for implementing exploration- and inquiry-based learning at child-care centres. Furthermore, recommendations are to be derived for future anchoring of the concept in the regular qualification offers of the Foundation and its eponymous professional development initiative "Haus der kleinen Forscher".

The Foundation draws on the findings of the studies to engage in systematic reflection on its existing educational offerings and the impact-oriented development of future professional development courses. The Foundation publishes the results of the scientific monitoring transparently in its publication series; all publications are also freely accessible on the website[4].

A Scientific Advisory Board supports the Foundation on research issues and in ensuring the Foundation's services are professionally sound. It is composed of independent academics from various disciplines and makes recommendations to the Executive Board and the Foundation Board. The members of the Board are high-profile experts from relevant disciplines and are appointed for three years respectively. From 2021 to 2023, the following members are as follows:

- Chairperson: Prof. Dr Mirjam Steffensky, University of Hamburg, Department of Educational Sciences
- Prof. Dr Yvonne Anders, University of Bamberg, Chair of Early Childhood Education and Upbringing
- Prof. Dr Nadine Bergner, Technische Universität Dresden, Didactics of Computer Science
- Prof. Dr Fabienne Becker-Stoll, State Institute for Early Childhood Research and Media Literacy (IFP), Munich
- Prof. Dr Wolfgang Böttcher, Münster University, Educational Science
- Prof. Dr Marcus Hasselhorn, German Institute for International Educational Research (DIPF), Frankfurt am Main, Department of Education and Development

---

[4] All results and publications in the area of scientific monitoring are available in PDF format at: www.haus-der-kleinen-forscher.de, under the heading "Scientific monitoring". All results of the external accompanying research are also published in this scientific publication series. An overview of the volumes published to date is available at www.haus-der-kleinen-forscher.de or at the end of this volume.

- Prof. Dr Bernhard Kalicki, German Youth Institute (DJI), Munich, Department of Children and Childcare, and Evangelische Hochschule Dresden, Chair of Early Childhood Education
- Prof. Dr Olaf Köller, Leibniz Institute for Science and Mathematics Education (IPN), Kiel, and Kiel University
- Prof. Dr Nina Kolleck, Leipzig University, Political Education and Educational Systems
- Prof. Dr Armin Lude, Ludwigsburg University of Education, Department of Biology, focus on Education for Sustainable Development
- Prof. Dr Jörg Ramseger, Freie Universität Berlin, Chair of Primary Level Education
- Prof. Dr Hans-Günther Roßbach, University of Bamberg, Chair of Elementary and Family Education
- Prof. Pia S. Schober, University of Tübingen, Chair of Sociology with a focus on Micro-Sociology/Dr Ludovica Gambaro, University of Tübingen, Department of Education and Family
- Prof. Dr Christian Wiesmüller, University of Education Karlsruhe, Department of Physics and Technical Education, and Deutsche Gesellschaft für Technische Bildung (DGTB), Ansbach
- Prof. Dr Bernd Wollring, University of Kassel, Didactics of Mathematics

## 2 Relevance of Early Computer Science Education

We encounter computer science almost everywhere in our everyday lives, but we are often unaware of it. Whether we are waiting for the digitally controlled traffic light to finally turn green, operating our smartphone or switching on the fully automated washing machine: information technology is always with us and so it also plays an important role in the lives of children. According to a study commissioned by the German Institute for Trust and Safety on the Internet ("Deutsches Institut für Vertrauen und Sicherheit im Internet" (DIVSI)), "the increasing digitalisation of everyday life is already firmly anchored in family life for young children – as a topic and in their concrete actions" (DIVSI, 2015, p. 131). However, growing up in a digitally influenced environment does not automatically make children and adolescents competent users of digital technologies, as a study on computer and information-related skills among young people shows (Eickelmann, 2015). According to this study, almost 30 percent of young people in Germany do not have sufficient computer and IT skills for successful participation in society.

As part of the National IT Summit, which took place on 16 and 17 November 2016 in Saarbrücken, the German Mathematicians Association issued the following appeal: "Not the mere use of digital media, but the understanding of their fundamentals is what is required for an effective digital transformation. […] The aim should be to teach basic skills that enable learners to use digital innovations in a mature manner".[5] As such, computer science education is increasingly shifting into focus as a social task and should be an integral part of basic general education in the future. As a result, mastery of elementary methods and tools of computer science receives a similar status as writing, reading and arithmetic. All children should be given the opportunity to receive early education in this field. This means giving children leeway to ask questions about digital media and the informatics systems on which they are based and to seek answers through exploration and inquiry.

The "Haus der kleinen Forscher" addresses these challenges and aims to strengthen children's educational opportunities in a core area of digital education by providing further education and training for early childhood educators and teachers at child-care centres, after-school care centres and primary schools. While there are more and more initiatives on digital media use, the Foundation

---

5 DMV (German Mathematical Society) press release on "Bildungsoffensive zur digitalen Wissensgesellschaft" (Education offensive for the digital knowledge society). Mitteilungen der Deutschen Mathematiker-Vereinigung (2017), 24(4), pp. 191-191. Retrieved March, 2022, from https://www.degruyter.com/document/doi/10.1515/dmvm-2016-0074/html

focuses on computer science education at child-care centre and primary school age when developing early computer science education. There are no nationwide offers for computer science education for children aged 3-10 in Germany to date. While science, mathematics and technology education has been making its way into child-care centres for some years, there is a gap with regard to computer science education,
which is constantly widening in the course of the digital transformation of society. The Foundation's goal in this area of education is to give children their first experience in the field of computer science in order to develop a basic understanding of informatics systems in the long term. Thus, in 2017, a professional development module on the topic of "computer science" was added to the Foundation's offer and expanded for primary school teaching in 2021. The aim is not to increase the use of digital media, but to understand the underlying concepts. The Foundation offers a series of courses for early childhood educators and primary school teachers to address the topic of computer science in their work with three- to ten-year-old children – also without using computers or tablets – and to implement this as has been commonly practised to date with everyday materials.

In order to meet the Foundation's high-quality standards, the content in the area of computer science education was also developed with professional and scientific support. Since 2015, the Foundation has therefore been in close contact with experts in the field of computer science education who critically accompany and advise on the development of the topic in specialist forums and expert meetings (see the following chapter).

## 3 Professional Basis for the Subject Area of "Computer Science"

All content offered by the "Haus der kleinen Forscher" Foundation is developed based on the current state of scientific research on the respective topic. As already described, there are hardly any professionally sound or even evaluated concepts and educational approaches for the primary or even elementary level in the field of computer science education in Germany[6]. For this reason, the Foundation has sought expert advice and support for the professional establishment and further development of the Foundation's offerings in the field of computer science even more intensively than before in the other STEM educational fields of mathematics, sciences and technology. For the first time, renowned experts from international institutions and initiatives were involved in the development of the subject area. In addition, the Foundation is in constant dialogue with relevant partners and other initiatives active in the field of computer science education.

As part of the professional groundwork, the Foundation initiated the working group "Zieldimensionen informatischer Bildung im Elementar- und Primarbereich"[7] (Goal dimensions of computer science education at the elementary and primary level) with experts from the field of computer science didactics and primary school pedagogy (Nadine Bergner, Hilde Köster, Johannes Magenheim, Kathrin Müller, Ralf Romeike, Ulrik Schroeder, Carsten Schulte). From 2015 to 2017, the working group developed an expert report in which they formulated theoretically sound goal dimensions within the framework of computer science education for children in child-care centres and primary schools, but also for early childhood educators and primary school teachers, and examined instruments for measuring them. In addition, the working group examined the criteria for successful achievement of these goals and thus for effective and efficient early computer education in practice. Compared to the fields of mathematics, sciences and technology, where much more research findings were already available, the expert group has done excellent pioneering work in this context.

In order to discuss the initial results of the working group as well as the current state of research on early computer science education with an expanded group of experts, the Foundation organised the first international expert forum "Early

---

[6] One of the few offerings is the "Experimentierkiste Informatik" – a computer-science education programme for children of pre-school and primary school age, which has been developed since 2015 by "Forschungsgruppe Elementarinformatik" (Elementary Informatics Research Group (FELI)) at the University of Bamberg (http://www.uni-bamberg.de/kogsys/feli)

[7] More information on the working group is available at www.haus-der-kleinen-forscher.de/de/wissenschaftliche-begleitung

Education in Computer Science" in Berlin in autumn 2015 with leading national and international experts from science and practice. Experts from Germany, Switzerland, the UK and Slovakia addressed the question of how computer science education for children in child-care centres and primary schools can succeed. There was intense debate covering subjects such as whether the active use of digital devices is indispensable for computer science education or whether initial computer science skills can also be developed without the use of digital devices. It was agreed that computer science education should start early in order to give children their first experience in this area and to develop a basic understanding of computer science (systems) in the long term.

In autumn 2016, the second international expert forum took place in Berlin, focusing on the implementation of the expert recommendations in the Foundation's offerings of computer science education. This time, in addition to leading experts from computer science didactics, representatives of national and international practical initiatives in early computer science education (e.g., New Zealand, Great Britain) were among the guests and shared their ideas and experiences. The Foundation presented its first practical ideas and experiences developed on the basis of the working group's recommendations and discussed them with the experts. The experts emphasised the necessity of ensuring that children could relate the subject matter to their day-to-day lives. They encouraged the Foundation to make it possible to implement practical ideas with everyday materials in computer science lessons as well. This means that children can gain their first experience of computer science education even without using digital devices. Furthermore, it became clear that there is very little experience in this area of education, especially in the elementary sector. The Foundation's initial offers were therefore to be accompanied scientifically.

The results of the working group and the Foundation's first practical ideas were presented and discussed at the 4th meeting of the Foundation's Scientific Advisory Board in October 2016 in Berlin. The members of the Advisory Board appreciated the detailed work of the working group and emphasised the need for research projects in view of the lack of empirical bases in the field of early computer science education. Furthermore, they welcomed the Foundation's work to date in the field of early computer science education, which builds on professional expertise and international experience.

Complementing the expertise on the goal dimensions of computer science education, Nadine Bergner and Kathrin Müller developed an expert recommendation in which they present a selection of informatics systems for children at child-care centres and primary schools and describe how and under what conditions these can be used in the elementary and primary sector.

The key findings of the subject-specific groundwork for early computer science education are published in the present volume. The contributions focus on goals and concepts for successful computer science education in the elementary and primary sector and form the basis for the further development of the content of the Foundation's offerings in the field of computer science.

# Summary of Key Findings

"Haus der kleinen Forscher" Foundation

The ninth volume of the series "Scientific Studies on the Work of the 'Haus der kleinen Forscher' Foundation" is dedicated to computer science education in the elementary and primary sector. It contains a comprehensive report prepared for the Foundation by computer science experts and forms the theoretical basis for the development of the Foundation's content in the field of computer science. In addition, an expert recommendation provides an overview of informatics systems for children at child-care centres and primary schools and their use in elementary and primary education.

In the first report "Goal dimensions of computer science education at the elementary and primary level", Nadine Bergner, Hilde Köster, Johannes Magenheim, Kathrin Müller, Ralf Romeike, Ulrik Schroeder and Carsten Schulte specify pedagogical and content-related goal dimensions for early computer science education. Due to the lack of theoretical and empirical research findings in this area, the authors oriented their derivation of subject-specific goal dimensions for the elementary and primary level at the standards for the junior secondary level, proposed by the Gesellschaft für Informatik (German Informatics Society), as well as existing international curricula for early computer science education. The analysis of these existing concepts prompted the expert team to introduce a new process domain 'Interacting and Exploring', which they compared to the GI educational standards, in order to emphasise the importance for the hands-on and explorative handling of informatics systems at child-care centres and primary schools. The derived goal dimensions are discussed both at the level of the children and at the level of the early childhood educators and primary school teachers from the elementary and primary sector.

For the children, the authors recommend the following goal dimensions:

- Motivation, interest and self-efficacy in dealing with computer science (systems)
- Computer science process domains
- Computer science content domains

In addition, at the level of the children, goal competencies that are based on the guiding criteria described are prioritised and represent the most important age-appropriate links between process and content domains.

At the level of early childhood educators and primary school teachers, the following goal dimensions are recommended:

- Motivation, interest and self-efficacy in computer science education
- Attitudes, mindsets and understanding of roles with regard to computer science education
- Computer science process domains
- Computer science content domains
- Computer science didactic competencies

Furthermore, the authors discuss the success criteria for effective and efficient early computer science education in practise. These refer, on the one hand, to the competencies and attitudes of the early childhood educators and primary school teachers and, on the other hand, to the institutional framework conditions. Successful engagement with computer science topics in everyday life requires, in particular, competencies in computer science didactics on the part of early childhood educators and primary school teachers. This includes, among other things, identifying and designing effective learning environments, selecting materials with high computer science potential that are adapted to the children's individual stage of development, and (maintaining) the children's motivation in teaching-learning situations. According to the authors, an important prerequisite for the acquisition of these subject-didactic competencies is the subjective attitude of the educators and their motivation with regard to early computer science education. The educator's own interest in computer science topics should also have a positive effect on the children's motivation.

Another essential factor for the successful implementation of computer science education at the elementary and primary level is to avoid false or distorted ideas about computer science. In this context, the team of experts also sees the cooperation of educational institutions with families and decision-makers as an important criterion for success in order to counter reservations about the topic. With regard to the resources available at educational institutions, the authors emphasise that computer science education is also possible and expedient without the use of digital devices, especially at child-care centres. Especially with increasing age, the use of informatics systems offers opportunities to promote motivation and learning experiences and to link the content taught with devices which are actually available in the children's everyday lives. The prerequisite for this is then, of course, appropriate technical equipment at the educational institutions.

Finally, the expert report underlines the necessity of developing instruments to record the defined goal dimensions. The authors list important criteria for the development of empirical measurement instruments to evaluate the implementation and impact of early computer science education in practice. They emphasise the importance of scientific support in the implementation of this field of education.

In the second contribution of the volume, Nadine Bergner and Kathrin Müller give a professional recommendation on informatics systems for children in child-care centres and primary schools and their use in elementary and primary education. Even if the use of such systems is not an indispensable prerequisite for the implementation of computer science education at child-care centres, after-school care centres and primary schools, they offer an additional opportunity to teach computer science content in early education. The authors first describe a selection of suitable informatics systems for the elementary and primary level and give an assessment of the age for which these devices are suitable, which previous experience is necessary, both on the part of the learning support staff and on the part of the children, and which learning goals can be pursued with the systems. In addition, they establish a relationship to the goal dimensions of computer science education.

The conclusion of this volume describes the implementation of the recommendations in the content of the "Haus der kleinen Forscher" Foundation and provides a look ahead to the Foundation's future work. Based on these recommendations, the Foundation has expanded its range of services to include the area of early computer science education and has developed a professional development concept as well as extensive materials for practical use.

# A   Goal Dimensions of Computer Science Education at the Elementary and Primary Level

Nadine Bergner, Hilde Köster, Johannes Magenheim, Kathrin Müller, Ralf Romeike, Ulrik Schroeder, Carsten Schulte

1  Potential of Computer Science Education
2  Foundation of Goals on the Children's Level
3  Goals at the Level of the Children
4  Goals for Early Childhood Educators and Primary School Teachers
5  Examples of Prioritised Competence Domains for Computer Science Education
6  Prerequisites for Successful Early Computer Science Education
7  Conclusion

# 1 Potential of Computer Science Education

Many children in Germany today grow up with great opportunities and prospects. There are many reasons for this, not least of which is a good education system. Another reason that has an impact on everyday life but is not yet reflected to the same extent in the education system at present is the changes brought about by digitalisation, which is currently taking hold of and reshaping all areas of life. This will lead to a further variation and multiplication of opportunities and also to new kinds of challenges. In a narrow sense, digitalisation implies the conversion of analogue (i.e., continuously variable and thus theoretically infinitely different) data into digital form, i.e., a form that can be mapped onto digits and thus processed by computers. This means that, in principle, data from all areas of life that can be digitalised can be captured, stored, processed, transmitted and disseminated by machines at low cost. The resulting enormous increase in available information creates far-reaching opportunities and challenges for society.

At present, it is still unclear how to best respond to these comprehensive changes. Based on theoretical considerations, a look at comparable countries and their approaches, as well as the analysis of various practical projects and the state of research in computer science didactics, this expert report examines what contribution computer science education can make to contemporary and sustainable education in child-care centres and primary schools[8].

From our perspective, which is primarily shaped by computer science didactics, the following is clear: New digital media are not just another phenomenon in the everyday life of our children, they rather represent a new and independent educational area – besides others listed in the curricula for child-care centres or the competence standards for social studies and science in primary education (see Berlin, 2014; Gesellschaft für Didaktik des Sachunterrichts, 2013). This area of education has yet to develop, which is why we are using the discussions from different areas (digital education, media education, computer science teaching in the upper school) to design an age-appropriate concept[9].

The central goal of this new educational field or new perspective (at least partly new, as it cannot be exhaustively described by the technical perspective (see below)) is to act independently and responsibly in a digitally shaped world. This requires knowledge of the basic functional principles and modes of action of digital technologies, as otherwise they can – in the truest sense of the word –

---

8  *This article is a translation of the German version, dated 2018.*

9  *The terms computer science, computer science education, media education and digital education are not very clearly defined in the general discussion. In the course of this exposition, an attempt will be made to unfold and disentangle the various terms as much as possible.*

only be used superficially, and only insufficiently opened up, (co-) designed and evaluated.

We start with the question: What actually is computer science?

## 1.1 What is Computer Science?

Computer science is part of the immediate reality of children's lives – not only in the form of computers or other (obviously) digital devices, such as smartphones, tablets, photo- and videocameras, television sets, music players etc., but also in the form of devices and machines that are not immediately recognisable as such, like washing machines, clocks, microwave ovens, traffic lights and cars.

Computer science can be found everywhere where

- **processes are automatically controlled and regulated** (traffic light control, train timetable, the route of the rubbish truck or the programme of a washing machine),
- **data are stored and output digitally** (camera, audio book),
- **data are transmitted** (mobile phone, television, radio) or
- **data are changed and calculated** (weather forecast, calculator, car navigation system…).

Computer science education builds on experiences with digital devices to open up possibilities for (critical) access (not to be equated with handling) – especially access beyond the understanding of given processes: adaptation, configuration, construction and design. Computer science education is therefore largely based on the principles and concepts – hidden behind the user interface, so to speak – that are needed to construct and describe the functioning of digital systems, and thus enable their effective and efficient design and use.

Computer science deals with information that represents a real-world phenomenon alongside substances (subject of chemistry) and energy (subject of physics).

## 1.2 Computer Science as a Science

The term computer science (Informatik) was originally defined in Germany in 1957 as an artificial word made up of information and automatic: "They [engineers in the USA and Germany] found that it was possible to perform numerical calculations with electrical circuits, and at a speed that had simply been unimaginable until then. This was the beginning of automatic information processing. We call it 'COMPUTER SCIENCE'" (Steinbuch, 1957, p. 171)[10].

To this day, computer science is defined as the science of automatic information processing, and according to the Duden dictionary, the German term "Informatik" – computer science – means (Claus & Schwill, 2006): "the science of systematic representation, storage, processing and transmission of information, especially automatic processing with the aid of digital computers".

In addition to its roots in mathematics and engineering, computer science also has its roots in the methods and issues of natural sciences and is also an empirical discipline. According to Tedre and Apiola, these traditional lines are intertwined, but can be clearly distinguished in sub-disciplines (Tedre & Apiola, 2013). The following three traditional lines are still recognisable today:

- from automation or engineering: construction of technical solutions
- from mathematics: formal structure, abstractions, study of algorithms and increasingly
- from the tradition of natural sciences: explaining the world – in the sense of explaining and studying the digital world

In this context, computer science is (also) deeply involved in the field of psychological and sociological questions, research of media usage and usability (user-friendliness): How do people deal with digital devices? How do they have to be designed in order to be usable? What are the effects and possibly also the undesirable consequences of the spread of use and how can we react to these?

What does this mean for computer science education, which is what we are talking about here? To answer this question, we will take a closer look at individual facets of computer science in the following. In particular, we will look at the perspective of construction.

## 1.3 Construction in Computer Science

From an engineering perspective, computer science is a **science of construction**, which, at its core, deals with questions around and about the construction of digital artefacts[11] – hardware and software. Such questions are quickly of general nature, e.g., whether all (computable) problems can be solved efficiently with infor-

---

10 Quotations from German publications have been translated into English.
11 Artefact (lat. ars, artis, "craft" and factum, "made") stands for man-made objects (Wikipedia, 2014).

matics systems, i.e., in a realistic period of time. (They are not – most encryption methods are based on this insight.) The problems are often very specific and have a relation to practice or people. For example, the question of how to best develop software in a team; many software projects require hundreds of person-years: How can work be distributed such that you do not have to wait hundreds of years for the dissemination of the software?

For education, however, it is not so much the questions about technical details that are interesting, but rather the fundamental questions about how this construction takes place in principle, what is actually being constructed and how this affects the life of the individual.

One goal of computer science is to develop efficient algorithms that can be executed on digital artefacts, to automate the processes and to transform the data. But what does this mean? Why is this a novel technological achievement on its own, unprecedented in the history of mankind, which actually affects and very often radically changes all areas of life?

Already in the first didactic treaties of computer science at the end of the 1960s, the new – digital – technologies were understood as "the principal conclusion of the history of technology" (Frank & Meyer, 1974, p. 592) of mankind. According to this view, man has delegated more and more functions to tools, i.e., to objects. A distinction is made between three phases (see Frank & Meyer, 1974)

Phase 1: Objectification of limbs or organs such as fist, teeth, hands by means of corresponding tools derived from the environment, which were usually modified for this purpose (stone, hand axe)

Phase 2: Objectification of physical labour through machines

Phase 3: Objectification of mental labour through computers

While the first ideas about computer science teaching – at a time before (!) the invention of the PC – focused on the hardware of this new technology, a consensus quickly developed that is still valid today in abstract form and which briefly summarises the essential ideas of the algorithm-oriented approach in the 1976 recommendations by the Gesellschaft für Informatik (GI) on the objectives and content of computer science education: "The subject of computer science education is not primarily the technical function of the computer. Rather, it seems essential to know and recognise the possibilities of using the computer, as well as the effects and limits of the use of computing devices" (Eickel et al., 1969, p. 35).

In 1976, the aim of exploring the "possibilities of using the computer" was understood to mean discarding the misconceptions about the computer, such as that it is an "electronic brain" and contains "button-pressing automation" and establishing a rational understanding. Specifically, this means: "understanding of the problem of the possibilities", "classification of computer science knowledge in the world of experience", as well as special computer science knowledge (Eickel et al., 1969, p. 35).

From a methodological point of view, the early curricula proposed a focus on programming. This means that learners should (in principle) acquire basic knowledge of computers and computer science concepts through their own experience of the construction processes, be able to assess the possibilities, limitations and implications, and finally use and apply what they have learned independently through algorithmic problem solving and automation.

In implementation, however, the focus of the lessons was – at least initially – too often on practising individual programming language constructs rather than on solving problems and reflecting on these problem solutions: Instead of training thinking and problem-solving skills, the commands of a programming language were presented and practised as isolated items. To prevent this, computer science education was subsequently designed to be very project-oriented and thus also – at least to some extent – interdisciplinary: The focus was to be on solving real-world problems of application. In addition, less emphasis was placed on the actual technical construction in order to concentrate on the actual planning and design processes (more detailed in Schulte, 2001). This orientation is also referred to as modelling, as opposed to programming (cf. Figure 2).

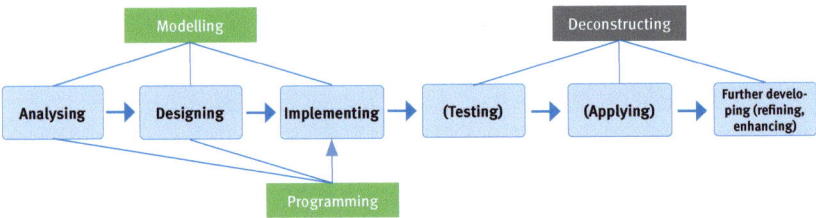

*Figure 2.* Steps in the construction process in computer science (boxes in the middle) and their respective emphasis: Programming emphasises implementation, modelling emphasises design. Deconstruction, as well as newer cyclical and agile models increasingly address the constant adaptation and evolution of existing products and sub-products.

In general, analysing is about understanding the problem and breaking it down into individually solvable sub-problems. A solution is then developed for these in the design phase and implemented in the implementation phase. These steps also apply to problems outside of computer science, so that the teaching of such general problem-solving competence is always seen as an important contribution of computer science education to general education. However, this does not yet explain the special significance of computer science-based problem-solving for the digital world. Therefore, these problem-solving and construction activities will be studied in more detail below.

In accordance with the division into different development stages of technology use mentioned above, the unique feature of computer science is that the automation of mental processes is in the foreground. Automation (of mental processes, or as we would rather say today: of information processing) is thus also the goal of computer science construction processes.

For modelling – especially also analysis and design steps – this means the following: The starting point is a specific (problem) situation. This must first be delimited and generalised with regard to a purpose because (only) generalised and thus recurring aspects of a situation stand a greater chance of being meaningfully automated. The solution can then also be used recurrently.

This generalisation of a situation therefore also means recognising and understanding the concept behind it. Modelling thus creates a model of the general process in the real world[12].

One can imagine this as follows (cf. Figure 3): Starting from S, model C is created (S stands for situation, C stands for generalised concept). This situation analysis is a deliberate and intentional abstraction to the essentials. The essential element depends on the desired purpose of the later solution and can always only be decided in the so-called socio-technical context – because purposes are created by and for people. Analysis and abstraction therefore also take place in the context of the development of an informatics system and in the context of clients, contractors and users and are not to be understood as neutral or value-free, but intentional.

---

[12] In this context, the real world also includes desired objects and situations that do not yet exist but are to be created.

Modelling S → C means: to abstract. Because abstracting is:

> a) generalising by reducing unimportant detail of the respective different situation to the decisive commonality.

At the same time, abstracting means

> b) retaining the important aspects with regard to the purpose of modelling

and also

> c) recognising the commonalities at a higher conceptual level: so to speak seeing the forest out of the trees.

*Figure 3. Definition of abstraction*

These steps are located in Figure 2 in the area on the left. In programming, these steps are called the analysis phase.

Based on the analysis, a solution can be designed (the step from C to C' in the figure). In this step, the model or parts of it are mathematised and formalised. This is the crucial process in designing in computer science: In the digital world, in the informatics system, all steps that are to run automatically must be clearly determined and described!

This model can then be implemented, so that a machine can execute it automatically and a user can apply it (in Figure 2).

The application of the solution is then again situation-specific, i.e., for example, with precise input data (in Figure 4, step C' to S'). By applying the implemented model, e.g., insights about the real world can be gained (simulation).

The complexity of such a process can be shown using the example of the 'calendar': If a date is to be advanced by one day, this can be done with the following input: day n → day n+1. This step seems simple at first. At the end of the month, however, the formula becomes more complicated: day n+1 is no longer sufficient here, but must be extended by month +1. At the end of the year, in leap years or with different month lengths, further information must be added respectively (in detail for this example Caspersen & Kolling, 2009). The example shows that in a computer science solution all steps and decisions must be considered and unambiguous. Therefore, even with simple examples, it quickly becomes quite complex and difficult to keep track of the individual steps and decisions.

Figure 4 provides an overview of the summary of the construction steps:

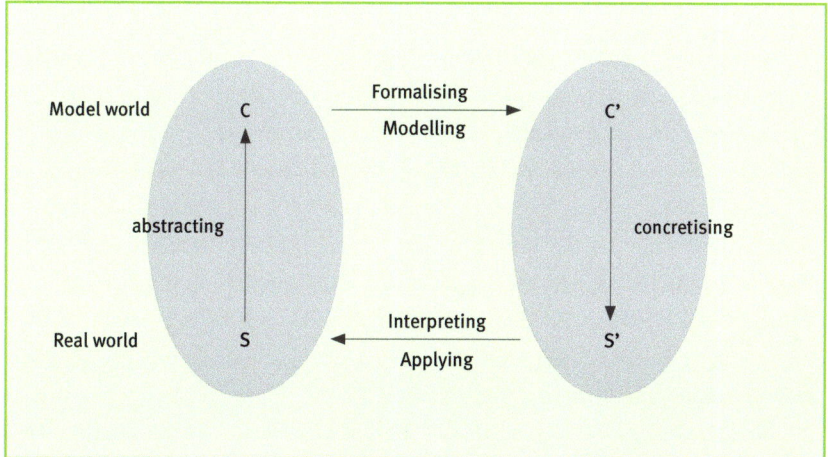

*Figure 4.* IT modelling (see Humbert & Puhlmann, 2004, p. 71; Schulte, 2003, p. 45, 72)

In summary, the computer science design process is thus structured as follows: From a situation or a single phenomenon (S) in the real world, a model is abstracted which describes the phenomenon as a concept (C). The model is then transferred to the informatics world or the model world (C') in mathematised and formalised form. Model C' can then be concretised by creating a specific programme from its abstract description (so-called specification). This programme is then executed on a computer. The application of the programme has an effect on the real world in turn.

What is decisive now is the difference between the real world and the model world. In the real world, the exact definition of an object or phenomenon is usually intuitively clear, but the precise demarcation is often difficult: For example, where does the mountain begin and the valley end? Describing ideal types is usually easy, but there are always blurs. E.g.: At what number do we call a collection of trees a forest? In the model world or the world of computer science, it must be possible to relate all these blurs to unambiguous statements. In this view, all individual situations and phenomena can be traced back to an unambiguous concept. This retracing is the basic prerequisite for digitalisation and automation. Peter Schefe describes this situation from the perspective of a software developer as follows: "The fundamental dilemma of software technology is to have to formally reconstruct what cannot be formalised" (Schefe, 1999). Thus, in computer science construction, an unambiguity is established, which is not always present. In development, decisions are made in this regard that are often relatively unimportant from a purely software-engineering point of view – for example, how big 'n' has to be to speak of a forest. From a user perspective, this can be quite relevant, for example, if owners have to pay a different tax rate for 'forest' than for a collection of trees.

So, the decision depends on external factors, on the intended use, or is simply made in the course of development. An important aspect is that these decisions are normative and have an effect. They define aspects that are later valid in the application and thus in turn shape the initial situation in reality (S becomes S').

Informatics solutions therefore always establish a uniqueness that may not have existed before. And: They always do this for a specific purpose, and this intended purpose entails another important consequence: A change in reality, an effect, is always intended. The automatic execution is intended to create or replace something – and thus the digital artefact used becomes part of the previously analysed reality (in the figure, the step from S' to S).

However, the model (Figure 4) falls short at a crucial point: The solution itself does not remain in the model world, but becomes part of the real world and changes the initial situation S. In principle, therefore, the solution always comes too late: As soon as it is used, the situation for which it was once developed no longer exists. This drives the cycle of changing software versions. Commercially distributed products are also usually offered not only by name but by name and version number, and new versions can be acquired in quick succession. A single software product can therefore only be understood by looking at the series of predecessors and possible successors. In general terms: The (further) development of programmes or digital systems and infrastructures takes place in a co-evolution: on the one hand, triggered by changing conditions and new ideas in the context of use, the social side, and on the other hand, equally driven by internal technical requirements and further developments. Digital systems or informatics systems are therefore increasingly understood as socio-technical systems (Magenheim, 2000).

With the increasing importance of computer science in everyday contexts, it is also becoming clearer that a purely inner-technical view is not sufficient to develop useful systems. And so, the theoretical view of the subject or field of computer science has also changed: In the past, it was easier to assume that clear and well-defined requirements could be placed on the construction of systems and that these requirements came from the outside, so to speak. Computer science itself concentrated on the technical implementation of these requirements – the path from C to C', i.e., also in software development or programming and the aspect of implementation. It dealt with what was then conceptualised as the purposeless construction of information-processing technologies. Its world was a mathematically describable world, defined by the possibilities and limitations of computers that could be fully described as a purely mathematical model (this is the basis of the mathematical view or tradition of computer science).

In this view, informatics systems could be considered with the paradigm of input-processing-output, focusing on the middle domain: Automation (hence the early coined term of the scientific discipline as informatics; see above). In the meantime, however, informatics systems are no longer adequately covered by this automation perspective, since interaction with distributed digital infrastructures such as social networks, flight booking systems, etc. produces much more complex processes.

Peter Wegener summarises these processes with the term **interaction** (Wegner, 1997). While in the purely or narrowly understood technical side of such systems, 'only' algorithmic processes can be observed, which can be recorded with corresponding mathematical description possibilities, the behaviour of the overall system emerges through parallel inputs from possibly millions of human users. In this way, however, the socio-technical system as a whole goes beyond the possible conditions that can be calculated comprehensively in advance.

All in all, it becomes clear that the essence of computer science can be determined by its construction process – and that computer science has a lot to do with model chains and different degrees of abstraction and modelling. Bernd Mahr even concludes: "However, no matter how you look at the science of computer science, models always play a dominant role" (Mahr, 2009, p. 228).

And further:

> "The close connection between computer science – in practice and science – with models also becomes clear when one considers the guiding question that explicitly or implicitly underlies its practical and scientific work. In computer science, this guiding question, which governs all engineering principles, presupposes the situational context of an information technology system development and reads: Does system S meet the requirements placed on its application? In computer science, there is probably no activity that can be taken seriously, that is not in some way in the context of a system development, be it that this development is only generally thought of or that it is specific. And to be taken seriously, this activity must contribute, directly or indirectly, to answering the guiding question it raises" (Mahr, 2009, p. 229).

Mahr continues with an analysis of modelling and the role of models. In Figure 4 in this chapter, we have described the modelling process. This can be supplemented with Mahr's analysis (see Figure 5 below):

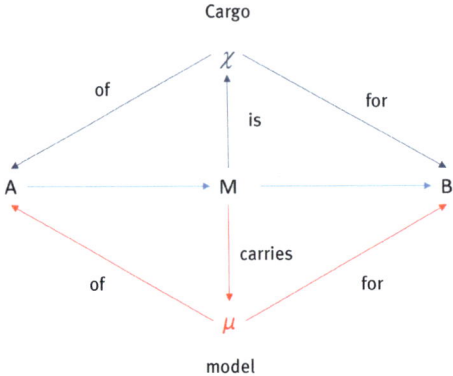

**Figure 5.** Structure of model relations by which an object M as model µ becomes the carrier of a cargo χ.

According to this, one cannot see per se whether an object is a model; instead, its "being a model" only proves itself in use (lower half of the diagram): Object M can be a model of A or (possibly without reference to an actually existing A) a model of B (or both). Mahr's decisive new view is now the function of being a model (upper half of the diagram): An object is (only) used as a model if it transports something, i.e., if it carries information that cannot be transmitted in any other way or only with difficulty. He refers to this information as cargo. When, in computer science or construction, we talk about abstraction, and this is considered difficult, then we can use this model to show that the difficulty lies in finding the appropriate cargo. And this makes it clear once again that modelling only ever becomes clear in relation to the purposes and the socio-technical embedding of the construction.

(Mahr's model is also interesting because it delineates the role of computer science or modelling in computer science in comparison to other disciplines – and because it is used in science didactic research, such as in the development of "model competence" in biology and physics.)

## 1.4 Similarities and Differences in Computer Science in Comparison …

### 1.4.1 … to the Natural Sciences

At first glance, the difference between computer science and the natural sciences can be easily summarised: While the natural sciences seek to investigate and explain the natural environment that surrounds us, i.e., the environment given by itself – nature, computer science looks at an artificial element: technical systems and procedures of automated information processing. Computer science and technology investigate the nature and design of artificial objects, which in this context are called artefacts ("lat. ars, artis 'craft' and factum 'made' stands for man-made objects", Wikipedia, 2014).

To put it more bluntly, one could say: The natural sciences deal with given phenomena of objects, while computer science and technology deal with artificial or man-made objects.

In both cases, therefore, the essence and the properties of these objects can be investigated: their internal structure, their mode of operation or principles of action etc. – the question of the **structure** of objects. However, there are two decisive differences in this respect:

1. Computer science as a science of construction poses this question mainly from the perspective of the developer and constructor: So, the answer is basically something like a description of the construction. Natural sciences ask this question to understand and explain. However, we believe that in educational contexts, digital artefacts should also be examined from this 'scientific-understanding' perspective in view of the increasing digitalisation of all areas of life – which children find to be just as much a given as the natural environment. In this perspective, 'computer science phenomena' are then examined analogously to natural phenomena. Here, computer science education – especially for younger children – will probably have to develop such explanatory models for digital artefacts even more as a subject didactic research task in the future, since a subject-scientific explanation is not necessarily the appropriate model of understanding for a constructor.
2. The study of **structure** alone, however, falls decidedly short and fails to recognise that we are not dealing with natural phenomena whose existence and structure must be taken as a given. Such a far-reaching transfer would denote what critics sometimes accuse computer-science education of: Children's adaptation to the digital world and its structure. Thus, if digital artefacts/phenomena are to be studied, the designed, the man-made element, must always be

included. For their structure has not developed naturally, but has been constructed with a specific intention. Therefore, it is important to always consider the **function** or intention of the artefact (Schulte, 2008a, 2008b). It is interesting to note that the same function could usually have been achieved by a different structure.

### 1.4.2 ... to Technology

In technology, expertise for the "Haus der kleinen Forscher" Foundation highlights the following characteristics for technology and technology science, which (at first) sound exactly like characteristics of computer science – especially in the constructive view adopted above. The characteristics are: "the design openness of technology, the value-bound nature of technology, the discursive character of the decision on the design of a specific artefact, the evaluation of the appropriateness of technical solutions, the culture-shaping effect" (Kosack, Jeretin-Kopf & Wiesmüller, 2015, p. 39). In the enumeration, the term technology can easily be replaced by computer science. But where then are the differences in the sciences and the specific contribution of computer science education? These differences become clear in the concretisation; for further on, the technology expert states:

> "Technology education must include:
>
> - aspects of the natural interdependencies that become effective in every specific artefact,
> - aspects of creative openness for specific problem solutions in connection with the actual activity, as well as
> - aspects of the evaluation of artefacts with regard to the functional fulfilment, the side effects and the embedding in a cultural context.
>
> These side effects must be concretised in terms of content, e.g.:
>
> - it must be possible to explore natural cause-effect relationships when using materials and tools insofar as they could be of significance for the purpose of the artefact. Teaching must provide space for experiencing basic laws such as the law of levers, Hooke's law,

Ohm's law etc., without the laws having to be formulated mathematically. Qualitative proportional relationships – especially in the lower grades – are quite sufficient" (Kosack et al., 2015, p. 39).

"Natural cause-effect relationships" of "material" look significantly different for digital artefacts than for technical ones. In particular, abstract material of "software" must first be made tangible. Let's therefore take a closer look at the step from Cn → C' in the diagram (see Figure 4): As formulated above, model (C) that emerged from the analysis and abstraction was made unambiguous through formalisation and mathematisation. But what "materiality" of the building substance forces this kind of formalisation and unambiguity? It is due to the way informatics systems process data: in simple, unambiguously defined steps. So, for example: Now the valley begins, so X applies. X could be defined as: We are on a bridge. In the technical construction of the bridge, on the other hand, there is usually more leeway as to where exactly it begins. The safety of a bridge can also be technically guaranteed in a different way: If the load-bearing capacity is perhaps not sufficient, the material thickness of the bridge piers can be increased somewhat. This is not possible in computer science: Here, it brings no additional gain if, for example, a case distinction is built into a programme a second time in order to have it checked twice. An informatics system is always in exactly one state: Either the test has resulted in state X or state Y – a second test does not improve the "viability" of the result.

This state concept causes another difference in "materiality": In "natural" cause-effect relationships, small deviations usually have small effects: If, for example, a bridge pier has turned out to be a tiny bit thinner, it does not make that much difference to the overall stability of the structure. But if the current state is "slightly off" in an informatics system, this simply leads to a different state – and can then be a fatal error state, regardless of the "initial state". The general law that small deviations usually have small effects does not apply here.

That is why there are no "analogy conclusions" in digital "materiality": In natural materials, similar compositions usually also behave similarly – in digital systems, similar initial states of a system can lead to completely different target states.

Reinhard Keil (Keil-Slawik, 1994) mentions further peculiarities of the materiality of software and comes to the conclusion that software should be regarded less as a "product", and more as a "collection of plans": In the programme text, developers determine "what is to be executed at the runtime of a programme, in what order and under what environmental conditions" (Keil-Slawik, 1994, p. 4). In doing so, it is important to "completely record all special cases and exceptional conditions and to process them appropriately" (Keil-Slawik, 1994, p. 5). According to Keil, with reference to Peter Naur (Naur, 1985), programming or software development can thus be understood as theory building: Software development should not(!)

> "primarily be understood as a production of programmes and associated texts, but rather as a process in which programmers develop a theory about how existing problems can be solved through programme execution. However, since not all problems and decisions occurring during system development can be documented with all their interrelationships, this theory exists only in the minds of the developers. Reconstructing the theory on the basis of documentation alone is completely impossible" (Keil-Slawik, 1994).

Mittermeir elegantly summarises this difference in materiality, and highlights another aspect that may certainly be very interesting for the introduction to computer science education:

> "In contrast to other technical subjects, in computer science, we do not construct by physical material processing, but by linguistic formulation" (Mittermeir, 2010, p. 59).

### 1.4.3 ... to Mathematics

Formally, computer science has a lot to do with mathematics: At universities, mathematics and computer science are often combined in a common department. Although historically incorrect, the term computer science is often understood to be composed of information + mathematics; the concept of algorithm originally comes from mathematics and is important in both disciplines. The GI standards (German Informatics Society) for the junior secondary level are modelled on the American mathematics standards. The concept of modelling (modelling and implementation instead of programming) is also found in mathematics.

From this perspective, it is difficult to clearly distinguish computer science from mathematics. In both disciplines, this view focuses on the problem-solving process, which essentially has a discipline-specific feature that the problem solution represents a formal model. From this (narrow) perception of computer science, Eberle, for example, argued as follows:

> "In this [=in the problem-solving process] a distinction must be made between the pure translation of colloquial language into formal notation (the level can vary) and the anticipatory understanding of temporal processes (procedures), in which their variables contents change (e.g.,

iterations). The latter requires additional cognitive effort and also distinguishes informational thinking from large parts of mathematical thinking (at least in relation to mathematics in the senior secondary level) that formalises static relationships" (Eberle, 1996, p. 329).

This reasoning refers to a rather subtle distinction of more static and dynamic aspects of algorithms, which distinguish computer science education from mathematics (see Claus, 1977).

Accordingly, computer science education, e.g., in the form of the first didactic approaches (algorithm orientation) and the purely mathematical tradition of computer science, was seen as a purposeless "inner-technological" consideration of the essential conditions of construction without reference to the reality of life. However, important aspects such as interaction (cf. "programming or being programmed" (Rushkoff, 2010)), reciprocity and co-evolution of action, individual and social unfolding and development in interaction with the technical infrastructure and its changes were lost. The system-building element was also missing. However, computer science – like technology – always raises the following question: How do we want to live?

Mathematics is neutral, so to speak, with regard to the goal of formalisation and mathematisation. Computer science, on the other hand, is always about automatic processing, about the (possible) implementation and use – and thus always about the referencing back of the solution S → S' in the figure above: The implementation (or more precisely the use of and interaction with) the solution in the initial situation, which the latter itself changes. The constructed system itself becomes part of the problem situation if it is used there and thus decisively changes the initial situation. The mathematical solution can be understood as an application in the sense of a descriptive, computer science application that uses an informatics system that did not yet exist in this form in the initial situation.

There are, however, also differences in the purely formal-abstract solution itself – since the physical properties of informatics systems on or with which the solution can be executed must be taken into account: Questions about the memory space required and time needed for executing the necessary computational steps change the character of the solution.

Furthermore, the construction of a solution is much more than the construction of an automatism because the latter must also be able to be used, tested and maintained. That is, *usability, maintainability* play a role – as do *extensibility* and *adaptability* (terms in italics are criteria for software quality) (see Wikipedia, 2016a).

There is another important difference that we consider crucial: In mathematics, formal problem-solving is represented "mathematically", i.e., in the (one) language of mathematics. Computer science makes use of mathematical expressions and representations – but distinguishes itself primarily by constantly developing new formal languages and representations that can be executed on automata or used to describe them. Interestingly, these languages are often also simple graphic notations (cf. Figure 6; for further details: Hubwieser, 2007; Hubwieser & Broy, 1997).

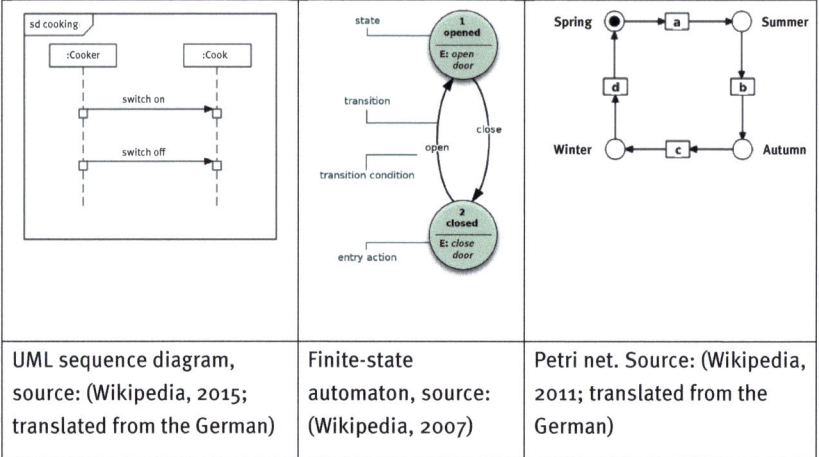

| UML sequence diagram, source: (Wikipedia, 2015; translated from the German) | Finite-state automaton, source: (Wikipedia, 2007) | Petri net. Source: (Wikipedia, 2011; translated from the German) |

*Figure 6. Examples for graphical notations*

These languages are often referred to as modelling languages because they are used to represent the result of a modelling process and are an intermediate step towards the implementation in a more machine-oriented programming language.

## 1.5 Computer Science and Computer Science Education

Now that computer science has been introduced as a discipline and distinguished from other sciences, computer science education can be characterised. To do this, we first draw on the results of a survey commissioned by the two major computer science associations, ACM and Informatics Europe, to clarify the role of computer science in education. One result of the report is the affirmation of the following distinction between "digital litaracy" (skills in using digital devices) and the introduction to the science of computer science[13]. A similar division is also found in the 'shut down or restart' report (The Royal Society, 2012) from the UK. The resulting

---

13  Both, "computer science" and "informatics" are commonly used for translating the German word "Informatik".

distinction between computer science, ICT and digital literacy is presented in Section 2.4.1.

In general, it can be said that almost all general education school curricula around the world teach both application skills and concept knowledge.

Application skills/application knowledge refer to the use of existing systems: "Any citizen of a modern country needs the skills to use IT and its devices intelligently. These skills, the modern complement to traditional language literacy in language (reading and writing) and basic mathematics, are called digital literacy" (Gander et al., 2013, p. 7). As examples of such skills, the report cites typing, creating and revising documents, searching, filing and retrieving data, handling data ethically and safely, selecting appropriate IT systems and the like (Gander et al., 2013, p. 8). Distinct from this are computer science and computer science education with their own topics and content, such as algorithms, data structures and abstraction (Gander et al., 2013, p. 9). The report sees the contributions of computer science education in particular in the promotion of creativity, the construction of artefacts, dealing with complexity and accuracy (Gander et al., 2013, p. 13). In the following, we will refer to such competencies as *computer science competencies*.

In this sense, digital education would be a juxtaposition of digital literacy and computer science education. This strict separation between user skills, on the one hand, and computer science education, on the other, has various causes. One of them is certainly the fear mentioned in the report of reducing computer science education to handling skills. Another reason is the desire to tie this new field of education to the – still quite young – academic discipline. In the German-speaking world, this is often linked to the statement that the discipline is called "Informatik" and not "Computerwissenschaft" (the "science about computers"), and thus has little or nothing to do with the artefact called computer.

Mittermeir argues, for example: "Informatik (computer science) is an artificial word composed of the fusion of information and automatic. The root word computer does not occur here. This means, that 'Informatik', although a technical subject, should not be a device-specific subject" (Mittermeir, 2010, p. 72). It is problematic if computer science education is reduced to the use of computers, giving a distorted image of the discipline (Mittermeir, 2010, p. 55). Computer science is "not really about the computer, but about constructed (i.e., technical) systems that allow data (in the broadest sense) to be interpreted in such a way that actions are triggered" (Mittermeir, 2010, p. 57).

Rechenberg disagrees with this view: It has been shown that "computer science is not concerned with information processes in society and nature". Therefore, one must "insist that computer science today is the science and technology of computers and their application, i.e., computer science" (Rechenberg, 2010, p. 47). Interestingly, however, the curricular proposals of the two are more similar than different. Brandhofer notes that it is no coincidence that the strict separation of computer science and computer (application) is hardly observed in practice and in curriculum discussions (Brandhofer, 2014, p. 3).

In our opinion, it makes more sense to start from the connection between the two supposedly separate areas. One way of doing this has been suggested by Puhlmann and Humbert. They start from phenomena in computer science and describe them in three categories (Humbert & Puhlmann, 2004):

1. Phenomena that are directly related to informatics systems: They occur during use and can contribute to making the application of the system easier, more efficient or even more enjoyable.
2. Phenomena that are indirectly related to informatics systems: They are not immediately recognisable, but only become apparent or perceptible during the analysis of the interaction with the system.
3. Phenomena that are independent of informatics systems: They are independent of digital systems and are characterised by having an inherent computer science structure and/or suggesting computer science thinking. Examples are phenomena in which search and sorting processes play a role.

In Humbert & Puhlmann, the use of or interaction with informatics systems is primarily intended as a trigger or pathway to the perception of computer science phenomena, which should subsequently be the focus of learning processes. However, both also write that the relevance or educational value lie precisely in the reference back to or in the application to the interaction. This approach leaves the question of whether 'Informatik' can or should be understood as a computer science. Accordingly, the authors define 'Informatik' somewhat vaguely as "the scientific discipline addressing the construction and design of informatics systems" (Humbert & Puhlmann, 2004).

Interesting in this expert document is the demonstrated relationship between computer science education and "application skills". The positive effect on skills in using informatics systems is repeatedly cited as a goal and rationale for computer science education (e.g., in Brandhofer, 2014; Gander et al., 2013; Humbert & Puhlmann, 2004; Mittermeir, 2010) – but this relationship is (mostly) seen only in one direction: While concepts and phenomena of computer science contribute to application skills, it is sometimes (implicitly) assumed that application itself is not part of computer science as a discipline. Therefore, application skills are not part of computer science or computer science education.

However, the seemingly simple and unambiguous separation between designing/constructing as a part of computer science, on the one hand, and the application as non-computer science, on the other, is not so clear or simple at all (Crutzen, 2000). Fischer et al. call for a corresponding rethinking in software technology, a meta-design that allows users to become designers, who participate in the adaptation, modification and further development of a digital artefact over its entire lifetime by (being able to) adapt(ing) and extend(ing) the respective artefact to their own needs – in other words, allowing the users themselves to become

designers (Fischer, Giaccardi, Ye, Sutcliffe & Mehandjiev, 2004). All in all, the postulated separation of "design-time" and "user-time" is becoming more and more obsolete (Maceli & Atwood, 2011). This applies not only to the (original) users, but also to the constructors: in the process of system design and development, it is also part of the tasks to examine already existing systems and to be able to competently apply the constantly evolving tools to support the various aspects of software development. Formulated in Crutzen's terms: applying and using is thus also an integral part of developing, i.e., of the core of computer science. In this perspective, using or interacting with digital artefacts takes on a new meaning that is fundamentally different from the idea of using tools that have been completed by others: Getting to know, familiarising, configuring and, above all, adapting, adjusting and extending are all part of interaction. And it also forms an integral part of construction, since this is not possible without the use of tools.

This leads to another important aspect, interaction includes exploration: Analysing and exploring digital artefacts with the goal of understanding them. Understanding includes being able to assess the possibilities and limits for adaptation and further development. And that, in turn, is not possible without reflection (e.g., with regard to the intended use and possible side effects) and evaluation. 'Understanding' here means in particular understanding the mode of construction because this is changed by the extension of the artefact.

Even if specific constructions and modes of construction are subject to constant further development, there is still a fairly stable core of computer science concepts that become visible or are anchored in these specific technologies and that can be taught. These can be called the 'fundamental ideas' of computer science (see Schubert & Schwill, 2011).

The concept of *computational thinking* has a different focus (Wing, 2006, 2008): It builds on computer science's knowledge of information-processing activities and uses computer science's techniques, models, concepts and tools for problem-solving thinking. At its core, Wing sees abstraction, and thinking in different layers of abstraction[14]. The aim here is to distinguish between what is needed to solve a problem and what can be left out. Complex problems are thereby broken down into parts and layers, so that a part can fall back on another or one layer on another. Thus, for example, an application programmer does not (always) need to know exactly how a command was programmed internally to use it. This has two important facets: First, sometimes the details of how something works internally matter – always when it comes to constraints, such as how much memory something requires and whether or not the available memory is sufficient. According to Wing, computational thinking is also about keeping an eye on the relationships between the levels of abstraction.

Fortunately, this thinking in abstractions can be supported by automation. Computing means "automation of abstraction" (Wing, 2008). Computational thinking deals with the question how a computer can solve a problem for me – for the answer or a good answer, the right abstractions and the appropriate "computer" must be chosen (Wing, 2008). It is important to note that "computer" in this context does not necessarily have to be a machine but can also be a human doing the "calculation" (Wing, 2008).

All in all, computational thinking is thus not just an approach or a way only of thinking, but an approach to dealing with problems that make use of computer science concepts and knowledge.

It is difficult to operationalise what exactly constitutes this computer science thinking at its core (National Research Council (U.S.) & Committee for the Workshops on Computational Thinking, 2010, 2011). In any case, it involves the application of concepts of computer science for problem-solving or in everyday life.

The suitable concepts of computer science can be derived – at least in part – from the development history of computer science education, provided that one can assume that the previous subject didactic concepts are not completely off the mark. In the early days of computer science education, there was – following the lines of tradition in computer science – the predominant mathematical or algorithm-oriented view of Turing machines[15], which is certainly also an essential foundation and which (has) concentrates(d) on the core area of algorithms. However, this was replaced quite early (late 1970s, early 1980s) by application orientation that sought to place not only algorithms but the significance of **algorithms** and **automation** for socially relevant areas at the centre of learning processes.

---

14  On the term of abstraction, see above Figure 3, p. 48.
15  A Turing machine is a universal automaton model proposed by the mathematician Alan Turing in 1936, which describes the functioning of a computer in a simple and thus easily analysable way.

This approach aimed to reconcile the mathematical view of the scientific discipline with the view of societal impact (Schulte, 2001). In school practice in academic secondary school lessons, this approach is still widespread (presumably because, to our knowledge, there is no empirical research on this).

Separate from this, partly due to political decisions (van Lück, 1986), was the introduction to **the use of digital systems** as basic computer science education, which was primarily geared towards application and use and, like computer science lessons, was intended in part to address the **interactions between computer science and society.**

Since then, the introduction to the use of digital systems in the school context has partly been experienced and understood as the opposite of computer science education: as pure user training that introduces the use of predetermined interaction paths. Under the title "Program or be programmed", Douglas Rushkoff (Rushkoff, 2010) describes this perspective, and at the same time an alternative: Either people adapt to these predetermined paths or they learn to design these interaction paths themselves.

In international comparison, computer science curricula often include the area of application/**interaction**. We have therefore decided to include this area as well, with the following emphasis: Interaction paths are designed by people and are thus also changeable. Even a 'normal' or 'simple' child-aged user, does not only have to follow the predetermined paths, but can and should also help to shape them (Gesellschaft für Didaktik des Sachunterrichts, 2013, p. 63). So for us, interaction does not mean "being programmed" but rather an introduction to active co-creation.

Before co-creating by changing the digital artefact, this competence also includes perceiving and recognising the (possibly different) interaction paths (e.g., for a task solution).

What is decisive for our perspective here is: Digital artefacts are man-made artefacts. Therefore, education can only succeed if the artificial, man-made element is taken into account. Children should realise at an early age that the digital world can and should be oriented towards human needs. They should understand its purpose-related nature to such an extent that they can judge whether the functions meet these needs or should be changed. They should experience themselves in the role of the constructor and be able to creatively help shape the digital world!

However, self-construction (or designing, changing, adapting…) presupposes that (some) basic computer science principles are known and that the users can assess, estimate and evaluate the relevance of their own construction and adaptation for their future actions. Even if they do not yet know any methods that allow them to familiarise themselves with a system in order to change it, children can already develop an understanding that these systems can in principle be designed: There are various approaches and systems with which children can be actively involved in the design (cf. sections 2.3.2, 2.3.3 and the examples in Chapter 5). If computer science education lays the foundation for this at an early stage, we believe it has fulfilled its mission to educate children of pre-school and primary school age to become mature and able to participate, also in the digital world.

In the following, we outline the process of constructing and exploring, as well as their interrelationship, based on the "inquiry cycle" method of the "Haus der kleinen Forscher" Foundation (HdkF, 2013a; Kramer & Rabe-Kleberg, 2011). The exploration cycle precedes the designing cycle – but designs that start directly without an example are also imaginable.

The two cycles alternate between two perspectives, which are explained by colours: Red refers to the construction or structure, green to the purpose or function. This duality is described in different variants and formulations in the didactics of computer science.

In the approach "Computer science in Context", the context can be seen, above all, as a reference to the benefit – the possibilities of use. Then, in the course of the lessons, decontextualisation takes place, in which the technical or inner perspective (red) is adopted. At the end, both are linked by recontextualisation (Koubek, Schulte, Schulze & Witten, 2009).

Peer Stechert developed a concept for "systematic exploration of the behaviour of computer science systems". In the sense of the black box principle, the externally visible behaviour (green) can be systematically analysed and finally the interplay of the externally visible behaviour with the internal structure (red) can be explored – in this step, the system is then analysed as a white box or glass box (Stechert, 2009).

The deconstruction of informatics systems (Magenheim, 2000) also starts from the analysis of "finished" informatics systems (IS). The crucial point is that various aspects of the IS can only be understood if the original design decisions of the developer are known. Certain features may be accidental or by-products, while others are deliberate design decisions or interpretations of the developers about the purpose of the system. However, since these can no longer be clearly reconstructed, they have to be recreated based on the existing traces. In this respect, it is more an interpretation than an analysis. This refers to the different views and materials that can be achieved: user interface, source code and source code comments, user manuals, developer documentation etc. So, in a sense, deconstruction attempts to derive the associated purposes and intentions from the documentation of the inner construction (red), e.g., the source code (green), and to expand the IS on this basis if necessary.

Duality reconstruction attempts to directly link the two areas red and green, i.e., inner structure and construction (red) and purpose, intention, use (green)

(Schulte, 2008b). One method is experimentation (Schulte, 2012).

Using a robotics experiment, Mioduser describes the presence of these two perspectives in 4-5 as well as 6-7-year-old children. He calls Str (red) the technical perspective and Fkt (green) the psychological perspective. In the psychological perspective, the children explain the robot's behaviour with psy-

chological categories, such as will, intention or personality – in the technological perspective, they explain the behaviour with causal cause-effect descriptions and rules or rule-governed behaviour (see Levy & Mioduser, 2008).

### 1.5.1 Exploration

Exploration of a given artefact (dA for digital artefact) or informatics system (IS) is similar to experimentation and thus somewhat similar to the inquiry cycle of the natural sciences. Here, the aim is to increasingly explore the internal structure, the individual components and their mechanisms of action on the basis of the externally perceivable function or intended use or on the basis of the use of the system – i.e., from the perspective of the user (see Figure 7 in this section). Through purely external observation, this can only be done to a certain extent, but most aspects of the user interface are designed to reflect the inner system states or aspects of them. The point is to draw the attention away from the task (writing a text, making the robot drive, taking a photo) to the technical implementation (text is not stored pictorially, but by characters and meta-characters that describe how the text looks on the screen…).

For this purpose, as shown in the figure, exploration can be carried out as a process with different steps, whereby attention is increasingly drawn from the function of the dA (shown in green) to the structure (one could also say to the mode of action or the principles of action). These aspects are marked in red. All in all, the aim is to achieve a connection between the function and mode of action, i.e., a balance between the two colours – only in this way can one understand that the mechanisms of action were not simply constructed in this way, but in relation to the intended use. Figure 8 shows the issues of the individual steps in more detail. Here, the reference to the Process Domains explained in the following chapters is already established.

One tool for this is the representation with suitable expressions. Computer science has developed various expressions and graphical modelling languages for the representation of informatics systems or individual aspects of informatics systems, which can be implemented here (cf. Figure 6 in this section, and in detail at Hubwieser, 2007).

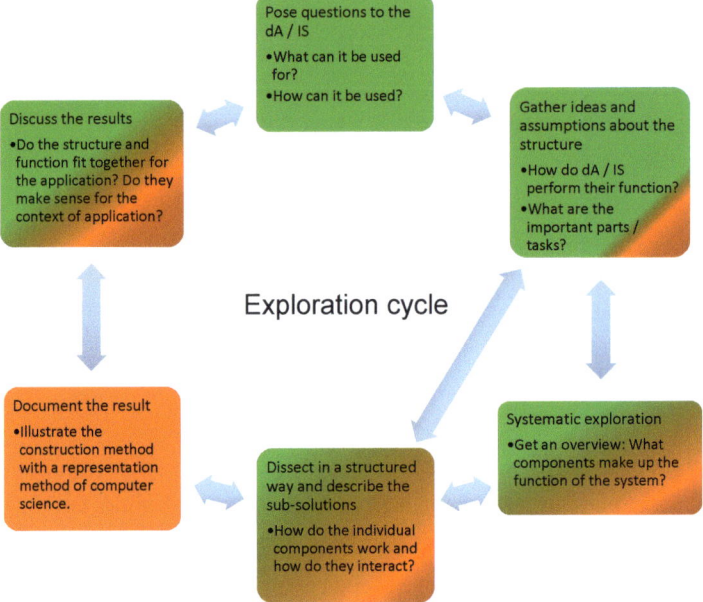

*Figure 7.* The exploration cycle for the investigation of digital artefacts (apps, digital camera, robot, text processing software, ...)

# 1 Potential of Computer Science Education

| Step | Pose questions to the dA / IS | Gather ideas and assumptions about the structure | Systematic exploration | Dissect in a structured way and describe the sub-solutions | Document the result | Discuss the results |
|---|---|---|---|---|---|---|
| Key question | What can it be used for? How can it be used? | How do dA / IS perform their function? What are the important parts / tasks? | Get an overview: What components make up the function of the system? | How do the individual components work and how do they interact? | Illustrate the construction method with a representation method of computer science. | Do the structure and function fit together for the application? Do they make sense for the context of application? Does the assumed structure match the actual behaviour? |
| Process domains | P0: Interacting and Exploring (low level) | P0: Interacting and Exploring P4: Communicating and Cooperating | P0: Interacting and Exploring P3: Structuring and Interrelating P4: Communicating and Cooperating | P3: Structuring and Interrelating | P5: Representing and Interpreting | P2: Reasoning and Evaluating |
| Perspective | External: Function | Primarily function, rough ideas, structure | From function to structure | Inside: Structure | Inside: structure with reference to function | Balance of structure and function |
| Explanations | First clarify what the artefact/system is there for and what it is needed for. In everyday use, perhaps only a limited area can be identified. Here, an overview of the area of application should be developed. This helps to dissect the dA and to develop ideas about internal working methods | In this step, ideas for internal working methods are developed, discussed and refined or already discarded. Initial linguistic descriptions for internal ways of working and concepts are also tested here. | Ideas are systematised and assigned to functional areas. General areas such as input, processing, output, storage … Guided by ideas; read operating instructions if necessary; examine from the point of view of operation | Data structures and algorithms, object interaction Stronger abstraction, decontextualisation and formalisation here; take notes here | Document the internal structure. Use case, state chart, flow chart, … | Contexts of use and roles of various stakeholders, check results in use, …. Can move into the design cycle here |

*Figure 8. Individual steps of the exploration cycle*

## 1.5.2 Constructing

In computer science, constructing refers to the creation of a digital artefact or an informatics system. The different steps basically start with the question of what is needed, and then the consideration of what has to be constructed to achieve this (cf. Figure 10). The steps are even more interlinked than in exploration, so that cycles, regressions, i.e., step-by-step approximation to the desired solution, are natural processes here. In addition to constructing a completely new system, it is also possible to build on the basis of an existing informatics system, which may need to be explored first. In the construction process, there can be local cycles, so to speak, both in the analysis of the required functions and in the design of partial solutions. In these local cycles, sub-problems are then processed one after the other (cf. Figure 9). Figure 10 shows the individual steps of a design cycle in detail.

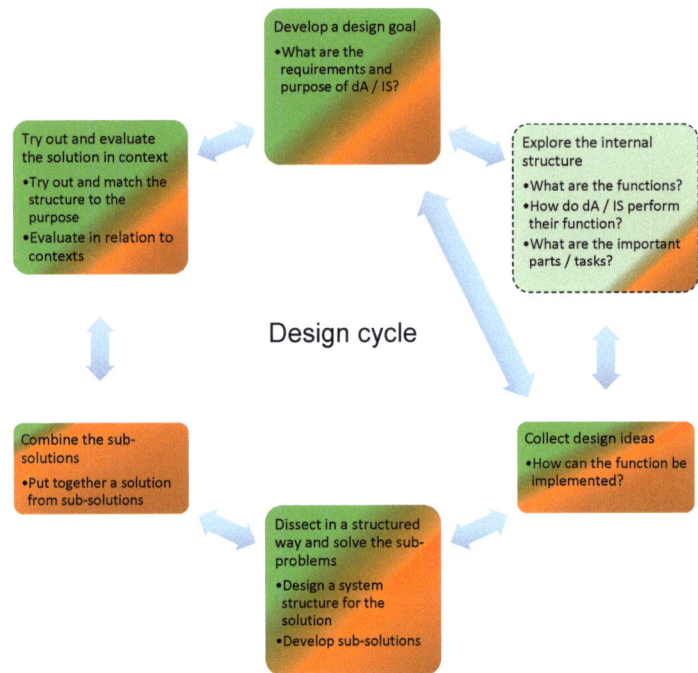

**Figure 9.** The design cycle for the construction of digital artefacts

| Step | Develop a design goal | Document the internal structure | Collect design ideas | Dissect in a structured way and describe the sub-solutions | Combine the sub-solutions | Try out and evaluate the solution in context |
|---|---|---|---|---|---|---|
| Key question | What are the requirements and purpose of dA / IS? | How do dA / IS perform their function? What are the important parts / tasks? | How can the function be implemented? | Design a system structure for the solution Develop sub-solutions | Put together a solution from sub-solutions | Try out and match the structure to the purpose Evaluate in relation to contexts |
| Process domains | P0: Interacting and Exploring (low level) | P4: Communicating and Cooperating | P4: Communicating and Cooperating | P3: Structuring and Interrelating | P1 Modelling and Implementing | P2: Reasoning and Evaluating P5: Representing and Interpreting |
| Perspective | Function with reference to structure | From function to structure | Function loop, especially structure | Function loop, especially structure | Structure | Balance of structure and function |
| Explanations | Develop an idea for the use of a system that does not exist yet. | Check the boundary conditions for the possibilities for realisation on the basis of the existing structure and function. Can move into the exploration loop here | Develop structural ideas for realising the function | Break down the structure and define it more precisely, also breaking down the function into sub-functions and developing individual structural solutions for them | Implementation phase | Use and evaluation phase |

*Figure 10. The individual steps of a design cycle*

## 1.6 The Relationship of Computer Science Education, Media Education & Digital Education

In the early discussions about computer science in education, computers (for computing – calculating) were often referred to as calculators; as we saw in the previous section, with close links to mathematics. However, computers were increasingly used in a wide variety of fields, including education, commerce, manufacturing, science, the military, health care or politics (see Magenheim & Schulte, 2006) – and as a result, the applications and effects of, as well as the view on computers changed. After a debate as to whether computers should be seen primarily as computers or rather as tools, instruments or media, the perspective changed at least to the extent that digital artefacts are now seen as interactive "media" in personal use and also mostly as networked online media.

With regard to the use of digital artefacts as "personal media", Keil characterises the "computer as a medium – media as a thinking tool of the mind". He states:

> "The view of the computer as an automaton that executes processes without human intervention, and the view of the computer as a tool, which, especially in interactive terms, gives the users scope for action and decision-making in structuring the process, is juxtaposed by Coy 1995 with the hitherto uncommon view of the computer as a medium. The support of distributed cooperative work processes was no longer compatible with the view of the tool, or could not be sufficiently justified with it, because: 'Networked cooperation is the basis of modern forms of production, based on the division of labour and is technically supported accordingly by networked computers: The computer becomes a medium. (Coy 1995a, 36)'" (Keil, 2012, p. 147).

These interactive media are thinking tools, since they enable feedback and thus the experience of difference. For example, by representing a calculation process, it can be perceived and checked for differences from the actual intended result: written thinking is thus more powerful than mental arithmetic. This support turns digital artefacts as personal media into thinking tools.

Computer scientist and thought leader Alan Kay goes so far as to consider digital artefacts as media through their combination of interaction and automated processing as a new form of cultural expression that replaces the book age.

In 1977, Goldberg and Kay described their idea of the Dynabook as follows: A device the size of a notebook "which could be owned by everyone and could have the power to handle virtually all of its owner's information-related needs. Towards this goal we have designed and built a communications system: the Smalltalk language, implemented on small computers we refer to as 'interim Dynabooks.' We are exploring the use of this system as a programming and problem solving tool; as an interactive memory for the storage and manipulation of data; as a text editor; and as a medium for expression through drawing, painting, animating pictures, and composing and generating music". (Kay & Goldberg, 1977, p. 1)

The Dynabook idea expresses exactly what Keil calls a thinking tool: An interactive tool for making and expressing thoughts. Expressing mainly in dynamic and interactive form. These media are thus not written nor programmed (in the sense of coded) in the conventional sense, but generated by an equally interactive tool in a heuristic writing and programming process. This creates a new form of media expression that only those who can programme know how to use. Kay and Goldberg have developed Squeak for this purpose, possibly the first object-oriented programming language.

The production of expressions is thus dependent on media in two ways: First, on media for dissemination and transport. Secondly, however, also for the interactive creation process. Media action is thus often, or even always, product-dependent and takes place, for example, in social networks: that is, in digital infrastructures provided by commercial companies. Media education points this out and likes to link this with the argument that a purely 'technical' view of this digital world is therefore not sufficient.

Döbeli summarises this discussion in the following figure:

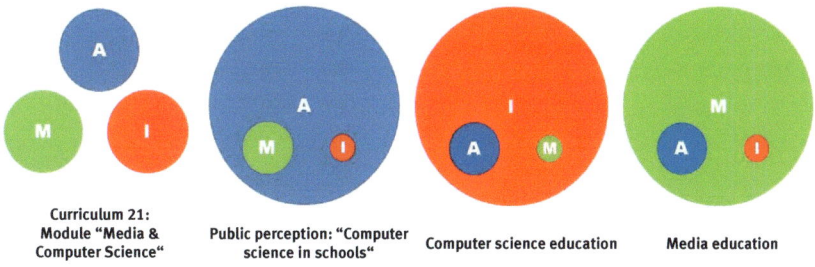

*Figure 11.* Different ways of conceptualising the connection between (a)pplication of IT (i)nformatics and (m)edia education (according to https://beat.doebe.li/talks/bern15/)

In the public perception, therefore, the applications are in the foreground: The efficient and effective use of digital artefacts or of computer science and communication technologies. From the perspective of computer science education, the focus is on computer science: Understanding the "basic concepts of automated information processing", using the basic concepts to solve problems and to understand information society. From a media education perspective, the focus is on media: Producing digital content, critical reflection of "use, meaning and effect" (according to: https://beat.doebe.li/talks/bern15/sld009.htm; Cf. also Döbeli Honegger, 2016).

The idea behind Figure 11 is to show that from the respective disciplinary perspective, the respective "own" cycle is seen as constitutive for the consideration of the other cycles. This is linked to the assumption that the respective "embedded" topics or competencies (from the other areas) are simply acquired along the way. In contrast, as with the attempt in Switzerland with curriculum 21 (Deutschschweizer Erziehungsdirektoren-Konferenz, 2016), it seems to make more sense to consider the areas more on an equal footing.

There are also attempts to argue less from one's own disciplinary point of view and instead describe competencies for the digital world in general or overarching terms: In 2010, an expert group of computer scientists and media scientists (commissioned by the Federal Ministry of Education and Research, BMBF) described competencies for the digital world. These are divided into four subject and topic areas:

1. Information and knowledge
2. Communication and cooperation
3. Search for identity and orientation
4. Digital realities and productive action

The individual competencies are then described quite abstractly without direct reference to the discipline. A competence in the field of information and knowledge is defined as follows: "understanding the production and dissemination of information and its exploitation as interactive processes and participating in an addressee-appropriate, situation-related and responsible manner" (Deutschschweizer Erziehungsdirektoren-Konferenz, 2016, p. 9).

Current approaches are looking for ways to reconcile the different aspects or views so that they can be fruitfully related to each other in educational processes. On the one hand, they aim to prevent what is often perceived as a pointless introduction to inner-technical principles of action without reference to the life-world and, on the other hand, to avoid operational training that only enables the use and thus adaptation of individual possibilities of action to given digital-technical systems. In this way, the socio-cultural perspective in particular should also be strengthened:

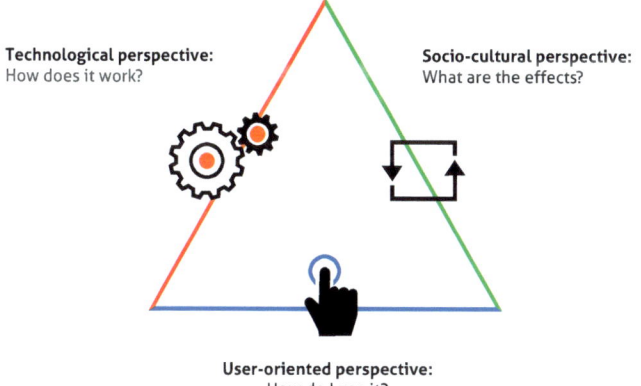

*Figure 12. The Dagstuhl triangle: Insight perspectives on the digital world (Brinda et al., 2016, english version by Beat Döbeli Honegger and Renate Salzmann)*

The Dagstuhl manifesto describes three perspectives in the following terms:

- The **technological perspective** questions and evaluates the functioning of systems that make up the digital world. It provides answers to questions about the operating principles of systems and answers to questions about their possibilities for expansion and design. It explains various phenomena with ever-recurring concepts. In doing so, it teaches basic problem-solving strategies and methods. It thus creates the technological foundations and background knowledge for co-shaping the digital world.

- The **social-cultural perspective** examines the interactions of the digital world with the individual and society. For example, it explores the questions: How do digital media affect the individual and society, how can one assess information, develop one's own points of view and influence social and technological developments? How can society and the individual help shape digital culture and cultivation?

- The **application-oriented perspective** focusses on targeted selection of systems and their effective and efficient use for the implementation of individual and cooperative projects. It addresses questions of how and why tools are selected and used. This requires an orientation about the existing possibilities and functional scopes of common tools in the respective application domain and their safe handling.

Following this manifesto, we present a model here that unfolds the perspectives shown in the Dagstuhl triangle. In this context, it is important that the individual perspectives intertwine. We also assume the digital, networked world in which children grow up today. The primary perspective for us is the technological one and thus the question: "How does this work?" Nevertheless, the other two perspectives on this issue of the digital world are equally important.

## 1.7 Conclusion: Computer Science Education for all

In accordance with the educational plans for the elementary sector and the curricula or the competence standards for social studies and science in primary education in the sense of basic education (Gesellschaft für Didaktik des Sachunterrichts, 2013, p. 9), we see the central task of computer science education in supporting children in "understanding the reality of their lives factually, in opening it up on this basis in an educationally effective way and in orienting themselves, participating and acting in it" (Gesellschaft für Didaktik des Sachunterrichts, 2013).

This is specifically about orientation in and understanding of the digital world, which is an essential part of the reality of children's lives today and in the future. It is also about getting to know new possibilities of expression and being able to participate creatively, participatively and responsibly in the further development of this digital world.

Within the framework of the educational goals for the elementary and primary level, we are essentially guided by the standards for the junior secondary level (GI – Gesellschaft für Informatik e.V., 2008) proposed by the *Gesellschaft für Informatik (German Computer Science Society)*, which also essentially feature the three perspectives explicated in the Dagstuhl triangle. However, we will extend these standards by one aspect: Interaction and exploration as a new process (P0). This is about application competencies, but not solely to the ability to use given specific systems. Rather, it is about the ability to familiarise oneself with unknown systems, to have general user competencies and to be able to apply them exactly, as well as to be able to think about interaction possibilities and consequences in an age-appropriate way. It includes aspects of reflecting, assessing and evaluating interaction possibilities and aims directly at the aforementioned general principles and participation possibilities in the digital world, as well as strengthening independence.

All in all, we thus want to include **interaction and exploration** as a separate area of competence. In contrast to application skills/digital literacy, which refers exclusively to the competent use of existing technological devices of the current generation (just as reading and writing refer to the currently existing national language and the current spelling rules), this is defined in an expanded understanding as a competence to explore and become aware of the underlying concepts and intentions of computer science. Exploring here means not only free play but also exploration guided by work assignments with the aim of "grasping" the systems, i.e., developing initial approaches to understand how they work (cf. sections 3.3.2 and 4.3.2, where the Process Domain is explained and discussed in more detail). Dealing with a digital artefact is thus only the base level of this area of competence. The focus is on system exploration, exploration through goal-directed interaction, which aims at understanding and acquiring and in particular includes aspects of adaptation and design in the sense of end-user design.

Computer science makes it possible to open up different ways of looking at a computer system or a digital artefact. For example, the concept of an automaton from theoretical computer science: A model that describes the functioning of digital artefacts as a set of fixed states that can be switched inbetween. However, these states do not all have to be displayed directly on the user interface. However, a corresponding change of state can be initiated and the meaning or effect investigated through targeted exploration.

Interaction as a competence refers not only to the operation of the system but also to exploring and reflecting on the meaning of the interaction. This meaning can go beyond the individual user and the individual use case and refer to a group of people, on a simple level e.g., in choosing a tool for one's toddler group.

It is open to what extent this is directly possible with children. However, they should not only learn to deal specifically with a digital artefact but also to confidently master general and transferable strategies for exploring an unknown system, while also thinking about its possibilities, limits and effects. An important aspect of this "exploratory competence" is the insight that most artefacts allow for adaptation and adjustment in relation to one's own wishes, i.e., they can be designed by oneself. This then leads to the possibility of being able to make these adjustments and to have the confidence to do so. It can be done through configuration, parametrisation and even smaller programming activities in the sense of end-user programming. In doing so, children can also realise that computer science is not only interaction, i.e., dealing with a system, but that it also involves designing and realising the interaction possibilities. This kind of programming competence is not only expedient for (professional) software developers but for all users of digital systems.

## 2 Foundation of Goals on the Children's Level

For the foundation of the goal of computer science teaching, the research situation and the experiences in the contexts of competence acquisition in the field of computer science are used. Based on the children's daily experiences with informatics systems, we will first look at the initial situation in Section 2.1, i.e., we will try to find out on the basis of studies if and how children are already using digital media, have gathered corresponding experiences, and where computer science education should start accordingly. In the following, we discuss the research situation with regard to the basic principles of learning psychology: What are the prerequisites for successful competence acquisition and how does interest in computer science develop? Section 2.3, outlines and systematises existing approaches to computer science education. The chapter concludes with a comparison of international curricula on computer science education and its classification in the chosen competence model.

### 2.1 Children in Digital Worlds

#### 2.1.1 Usage Experience

Children today gain experience with various digital artefacts, such as computers, tablets or smartphones at an early age: They observe how adults or other children use these devices, and they also use them themselves. The research on the experience of use of digital artefacts by children and toddlers has only just begun. However, the current studies do not take a computer science perspective but rather examine the experiences of users with regard to media education and media pedagogical aspects, and also with regard to the opportunities and dangers the digital world poses for children.

Since regular use of digital artefacts is not yet commonplace in schools and child-care centres, the responsibility for such usage experiences of children lies largely with parents or legal guardians. As the DIVSI U9 study has shown, the influence in this context depends on the perspectives and attitudes of parents, who differ significantly in their approach to and beliefs about the digital world, depending on their milieu (DIVSI, 2015). Chaudron's study comes to similar conclusions: According to this, children's usage experiences are "shaped partly by their parents' beliefs, values and ethno-theories" (Chaudron, 2015, p. 42). Here, it cannot be assumed that education-oriented homes are particularly predestined for frequent and high-quality usage experiences among children. On the contrary: These parents in particular are unsettled and worried by the negative and dramatised media coverage, especially with regard to data protection and privacy. In contrast, parents from more educationally disengaged backgrounds often see no need to actively intervene in their children's use of digital artefacts or to exert a regulatory or educational influence (Chaudron, 2015).

The use itself often takes place in isolation and unobserved – unlike most other child activities, which are much more often carried out together with the parents, siblings or other family members:

> "New (online) technologies were not perceived as an integral part of shared family life in most families. Rather, engaging with a digital device was considered as an individual activity, unlike offline family activities such as going to the park or playing a board game" (Chaudron, 2015, p. 28).

Chaudron also makes the point that many parents overestimate their children's 'digital skills' while underestimating the amount of time their children spend with digital artefacts. The associated assumption seems plausible: "This might be because so much digital use was to fill time when the parent was otherwise engaged" (Chaudron, 2015, p. 30). Children often gain access to artefacts precisely when their guardians are distracted, i.e., to bridge the children's waiting time and keep the child occupied when the guardians do not have time.

The extent to which parents grant their children use of digital artefacts or monitor such use certainly also depends on their own technical knowledge and skills. In basic studies on media use by children and adolescents conducted by the Medienpädagogischer Forschungsverbund Südwest (Media Education Research Network Südwest)[16] (MPFS – Medienpädagogischer Forschungsverbund Südwest, 2014a), setting up protective measures or access restrictions to online offers (filter parental control, limitation of usage times by software, router settings, pre-selection and installation of apps suitable for children on smartphones and other artefacts, etc.) is discussed – usually as the only example of specific technical or, in the broadest sense, computer science knowledge. In some cases, parents'

---

16 The results of regular studies (JIM study, KIM study miniKIM and FIM study) are published on https://www.mpfs.de.

quite justified fears have so far often led to little consequence in terms of their actions or behaviour (MPFS – Medienpädagogischer Forschungsverbund Südwest, 2014a, p. 73). All in all, the studies portray a picture according to which parents are rather overburdened with the shaping of children's usage experiences or the shaping of children's socialisation in the digital world (although with great differences depending on the milieu). It is partly recommended that child-care centres and schools should take a pro-active approach in order to help parents and also to intervene in a pedagogical and educational way themselves.

Currently available studies that examine the experience and behaviour of (young) children in the digital world paint a relatively uniform picture of user experiences in dealing with the "computer": The 2014 miniKim study states: "Computers and the Internet play a very subordinate role in the everyday lives of two to five-year-olds. At 85 percent, the clear majority has not yet had any experience with computers. Only every tenth child uses the computer – alone or together with the parents – at least once a week. In this case, almost twice as many boys (13%) as girls (7%) use the PC at least once a week. Among four- to five-year-olds, one in four children (24%) have had experience with computers, while computer use is the absolute exception among two- to three-year-olds (6%)" (MPFS – Medienpädagogischer Forschungsverbund Südwest, 2014b, p. 23). However, when taking into account that smartphones, tablets and other digital media are increasingly being used instead of "personal computers", the picture changes. For example, the DIVSI U9 study shows that 23 percent of 3-year-old children are already using smartphones and, from the parents' point of view, are "quite adept and intuitive in learning to operate end devices with touch screens" (DIVSI, 2015). A further significant increase of early use of digital media by children can be expected in the future. Thus, a European comparison in a study of four European countries shows: "children under five are heavy users of a number of digital technologies at home" (Palaiologou, 2016, p. 2). From an overview of different studies, Palaiologou concludes that most three to four-year-old children have acquired a kind of early "digital literacy", which helps them understand and use various digital technologies to obtain information and further develop first skills in dealing with printed language ("print literature") (Palaiologou, 2016, p. 6).

According to Plowman et. al., by using digital technologies at home, pre-schoolers can learn in four domains (Plowman, Stevenson, Stephen & McPake, 2012):

1. Manual operating skills in the narrower sense (e.g., in handling the mouse or gestures on touch screens etc.) although this domain is considered subordinate.

2. Acquisition of knowledge and skills in various domains, such as language and mathematics, especially via content in digital toys and media. This is the domain that the gaming industry highlights in its advertising.
3. Enhancement of social-emotional and cognitive skills, for example, through a sense of achievement during use.
4. Understanding of the role of technology in everyday life through own use and observing family members using technology. The authors present a case study to show that these learning opportunities depend not so much on the availability of technology at home, but rather on specific family practices and attitudes of parents (Plowman et al., 2012, p. 36). This also shows that, for example, the socio-economic status of the family is not a general predictor of the type of family interaction with technology.

Rosen et al. asked parents about negative effects. According to the study, increased "technology use" (TV, Internet, video games) leads to impairments (Rosen et al., 2014). On the other hand, Palaiologou states: "Analysis of the qualitative data emerging from the interviews suggested that children are ‚digitally fluent from a very young age'. One of the key findings was that parents felt that their definition of an illiterate person no longer corresponded to the traditional view of someone who cannot read and write, but rather was considered as a person who cannot learn, unlearn, relearn and use digital technologies as part of their everyday lives" (Palaiologou, 2016, p. 2).

In a summary, Gutnick et al. conclude that television still remains a leading medium (Gutnick et al., 2011; see also MPFS – Medienpädagogischer Forschungsverbund Südwest, 2014a, p. 73). However, they note a change around the age of about 8, when children increasingly explore digital and especially mobile media. Above all, mobile use or mobile media exert the greatest fascination. Palaiologou comes to similar conclusions or confirms this conclusion (Palaiologou, 2016).

### 2.1.2 Children use Different Digital Artefacts Confidently

According to the miniKim study, children up to the age of five mainly use games. "In second place are painting or drawing on the computer, slightly fewer boys and girls use a special learning programme. According to the main educators, a few of the two to five-year-old computer users already write texts or words on the PC" (MPFS – Medienpädagogischer Forschungsverbund Südwest, 2014b, p. 23). Palaiologou (2016, p. 11) found similar results.

The KIM study makes it clear that children aged 6-13 use different offers and apps:

> "The apps that users engage in at least once a week (regardless of where they use them) primarily include writing words or texts (50%), while a slightly lower proportion (45%) look at photos and/or videos with this intensity. A quarter paint or draw – girls (30%) more than boys (21%). With increasing age, writing (6-7 years: 27%, 8-9 years: 35%, 10-11 years: 51%, 12-13 years: 66%) and viewing pictures or videos gain importance (6-7 years: 31%, 8-9 years: 37%, 10-11 years: 44%, 12-13 years: 57%), painting and drawing, on the other hand, loses in attractiveness (6-7 years: 45%, 8-9 years: 34%, 10-11 years: 20%, 12-13 years: 16%). Of those aged ten and older, just under a quarter edit pictures and videos at least once a week – girls (27%) somewhat more frequently than boys (19%), the older ones more (27%) than the younger ones (19%)" (MPFS – Medienpädagogischer Forschungsverbund Südwest, 2014a, p. 34).

Some studies emphasise that e.g., due to the narrow range and repetitive nature of use, the activity observed in each case is carried out competently and confidently, but this is not an indicator of general competence in use or even an understanding of underlying internal processes or of functional principles: The KIM study of 2014 states in this respect:

> "When differentiating by age, it becomes clear that children's technical competencies that go beyond playing a DVD, usually do not develop until the age of ten. One third of eight and nine-year-olds can access the Internet independently without difficulty and one in four can print something out, but only just under a fifth can download songs to an MP3 player, while only one in ten is familiar with the filing structure of a computer. Among ten to eleven-year-olds, most tasks are only mastered by every third or fourth child. And even among the oldest children, there is a need to catch up in some activities" (MPFS – Medienpädagogischer Forschungsverbund Südwest, 2014a, p. 60).

However, the initial situation is changing slowly, but steadily:

> "The long-term comparison shows that technical competence has only partially increased in recent years, despite the omnipresence of media in children's everyday lives. Compared to 2010, significantly more children can access the Internet independently today. A clear boost in development was observed in the downloading of mobile-phone pictures. However, the ability to print things out, download songs to an MP3 player or set up a file system has developed just as little in a positive direction

as downloading files from the Internet" (MPFS – Medienpädagogischer Forschungsverbund Südwest, 2014a, p. 61).

### 2.1.3 Use of Online Offers by Children

When it comes to, for example, the ability to search for online offers, children (depending on their age) seem to prefer to follow suggested offers. This "search strategy" can be called 'browsing', when, for example, they watch suggested follow-up films on YouTube.

Online, children mainly use search engines and video portals like YouTube. In doing so, they often cannot recognise or distinguish if they are offline or online, when watching a video. They also use social apps and services quite often; especially WhatsApp.

In addition, they use Internet search engines, especially Google or fragFINN. They tend to use visual suggestions (e.g., on YouTube), instead of typing in search terms. Similarly, they can navigate the web, by following visually highlighted links, for example.

Parents often express concern about children being exposed to content that is not age-appropriate (MPFS – Medienpädagogischer Forschungsverbund Südwest, 2014a). Parents and researchers sometimes also express concern about the resulting digital footprint and privacy protection.

What is interesting here is that these concerns relate to explicitly created data, but not to connection data or metadata. Chaudron also notes that parents lack knowledge about online risks (Chaudron, 2015, p. 5). All in all, it seems unclear how to respond to/ how one should respond to children's "digital footprint" through their online use (Chaudron, 2015, p. 7). According to these studies, the role of educational institutions in this issue is also unclear.

Technical solutions are sometimes mentioned for the protection against non age-appropriate content or to restrict access times, however not all parents can implement them (see Chaudron, 2015, p. 42).

### 2.1.4 Summary

All in all, children (and parents) use digital artefacts confidently. This type of use gives the impression that they are generally familiar with the digital world and possibly also understand the essential principles and modes of action. However, according to the studies examined, the actual range of use is low, so that the observed confident use is also rather a side effect of the milieu-specific (limited) perception of the overall available possibilities of use. This applies all the more to consequences and effects of interaction that cannot be directly assigned to the

WYSIWYG[17] paradigm. These are then less perceived and remain unconsidered in one's own usage behaviour. Although, for example, concerns about data protection and privacy are quite widespread overall, the role of meta-data and connection data was not even recognised by the authors of the studies mentioned.

The increasing popularity of smartphones and tablets with touch controls as well as the increasing use of voice commands have already noticeably changed children's usage habits. However, the consequences for this are not mentioned. With regard to smartphone-oriented operating systems, however, it can be assumed that with these, the skills for organising and managing data, for example, are trained less frequently, since the operating systems usually complicate or completely prevent access to the file system. This could explain the contradictory interpretations mentioned above, e.g., whether children can organise access to photos they have taken themselves with their smartphone; e.g., the possibility to view photos from the mobile phone on the PC or TV. This is often surprisingly easy when photos are shared via commercial cloud services. However, if this private data is to be transferred directly without third parties and then be easily retrievable, this requires more understanding of technical possibilities and the set-up of appropriate operating systems.

All in all, a wide range of utilisation abilities has been observed. This also applies to the parents who can then assist their children more or less helpfully, depending on the situation.

## 2.2 Foundations of Learning Psychology

### 2.2.1 Cognitive Prerequisites

Computational mindsets and acting in relation to children is a research area that has so far been neglected (see Borowski, Diethelm & Mesaros, 2010). Learning environments that enable children to develop a basic understanding of how computers work or even develop creative and design potential, offer initial approaches to making the learning subject of 'computer science' accessible to children (Borowski & Diethelm, 2009; Borowski et al., 2010). Until a few years ago, however, the focus of current efforts for computer science education in childhood was on media literacy (Borowski et al., 2010).

Despite the overall still unsatisfactory research situation, findings from mathematics-related studies (Benz et al., 2017) as well as results from learning and developmental psychology can also be used with regard to central computer-science-related competencies.

---

17 WYSIWIG: What-You-See-Is-What-You-Get: Users immediately receive the visual result of the interaction (see also "WYSIWYG", 2016b)

In order to be able to describe the learners' cognitive prerequisites, a distinction is made between novices and experts, which can exist at different age levels (see Stern, 2002, p. 29). Child experts can even be cognitively superior to adult novices (Stern, 2002). Empirical studies on domain-specific knowledge shows that children can perform better on familiar subject matter and produce a more effective organisational strategy than adults who are unfamiliar with content (Sodian, 2002). It is assumed that younger children are often still inferior to older children, because they have had less time and opportunity to acquire knowledge and skills and that they often fail to solve tasks adequately, only because they still lack the necessary domain-specific knowledge (see Sodian, Koerber & Thoermer, 2006). Using the 'football' example, Schneider, Körkel and Weinert were able to show that young children who have domain-specific prior knowledge have an advantage over both more intelligent and older children in terms of understanding and retaining relevant content (Schneider, Körkel & Weinert, 1989). This would have to be verified for computer science task examples.

Based on initial experiences with computer science-related learning environments for pre-school and primary school children as well as some study results, there are already specific indications that children of pre-school and primary school age can acquire computer science-related competencies. With the help of study results, Schwill considered to what extent younger children are able to understand or apply fundamental ideas of computer science, such as 'recursion', the 'greedy method' (successive build-up of partial solutions to the overall solution), 'structured breakdown' and 'reproduction of hierarchical structures' (Schwill, 2001). From his theoretical observations, he concludes that "children as young as primary school age can grasp a number of important fundamental ideas in computer science, provided that the topics are prepared in an age-appropriate way and taught in lessons that take into account children's cognitive structures and are supported by actions or real objects" (Schwill, 2001, p. 17). Gibson found that even primary school children, who are not literate yet, can learn how to use graphical algorithms (Gibson, 2012). Understanding algorithms is regarded as central for computer science education (see Modrow, 2010; Schwill, 1995).

In contrast, the authors of a study in which children were asked to solve programming tasks using the graphical programming language "PiktoMir" found that children's understanding of computer science concepts was limited (Rogozhkina & Kushnirenko, 2011). Of the six children under the age of six, only two were able to solve all the tasks. The authors conclude that children in this age group are not yet capable of adequately understanding computer science concepts. Considering the age of the children, the result can indeed be seen in a positive light: After all, one third of the under-six-year-olds already manage to complete all tasks correctly! It should be investigated whether the result can be improved even more if the children can draw on earlier experiences. Furthermore, the results of such studies are closely linked to the programming environment chosen. For example, Portelance, Strawhacker and Bers report that children as young as 5 years are able to develop algorithms when creating games and animations with the ScratchJr programming environment, which has especially been developed for this age group (Portelance, Strawhacker & Bers, 2016). Weintrop and Wilensky thus proved that the graphical representation of algorithmic structures has advantages in learning compared to the textual representation (Weintrop & Wilensky, 2015).

### 2.2.2 Interest in Computer Science

With regard to the genesis of interest in computer science, primary school children are a promising target group. Children and adolescents are interested in new media and future technologies. This can be concluded from studies, such as KIM and JIM, which examine which media are used by which children and adolescents (MPFS – Medienpädagogischer Forschungsverbund Südwest, 2014a, 2015). The rapid increase in the number of media used, even by younger children, shows that there is a great deal of interest, at least in terms of the use of media. However, it is still largely unknown whether children and young people are also interested in the background of computer science. However, the enthusiasm that children show when they see little robots playing football or when they are allowed to programme themselves, e.g., in the "Roberta" project, can perhaps be taken as a sign that interest can be triggered (Petersen, Theidig, Bördig, Leimbach & Flintrop, 2007).

In a qualitative study, Yardi and Bruckman investigated the perceptions of 'computing' or the attitudes towards computer science among children and adolescents aged eleven and over in comparison to computer science students (Yardi & Bruckman, 2007). They found that among the teenagers surveyed, the prevailing opinion was that computer science was a boring, remote subject with no connection to the real world. In contrast, the computer science students were enthusiastic and fascinated by the possibilities: "[...] graduate students described their research as exciting, social, and having a direct and meaningful impact on the world around them" (Yardi & Bruckman, 2007, p. 39).

## 2.3 Access to Computer Science for Children

As observed in the previous section, research related to computer science education is increasingly targeting adolescents and younger children. In line with this research interest, initiatives and projects are emerging that provide in-school and out-of-school learning opportunities to engage with computer science. Such offers pursue three main goals:

1. to trigger interest in computer science at an early age
2. to convey a basic knowledge of phenomena in a world shaped by computer science (digital society)
3. to promote computational thinking

Different approaches focus on different motivations for dealing with computer science. They differ mainly in whether computer/informatics systems are used as tools in the learning process, or whether access is limited to non-computer-science technical learning material ("unplugged"). In terms of computer use, a rough distinction can be made between access via software on universal computers (including e.g., tablets) and access via programmable toys, i.e., special computer systems for children. Another access is presented, namely out-of-school learning places and communities that use computers or unplugged access but take up the common, mostly collaborative experience as motivation and guideline of learning.

### 2.3.1 Access Without Computer

Computer science as a discipline offers concepts, methods and ways of working that can be found in various forms in teaching-learning materials even without direct reference to computer systems. In contrast to many software-based tools for introducing computer science, the aim of the so-called "unplugged" materials is to help children understand the ideas and strategies underlying problem-solving in computer science, thereby explaining the phenomena of the digital world. The common feature of all materials/accesses is that no computer is necessary for the execution. At the same time, it is made clear in the materials that the computational mindset presented comes from everyday thinking (cf. also "computational thinking"). Access is mostly playful and action-oriented and aims to reach children and young people of different age groups. In recent years, several books have been published that try to reach children from toddler age with a similar approach.

In addition to these materials originating from the context of computer science, approaches from neighbouring disciplines, such as early mathematical education, also promote competencies that can be assigned to early computer science education.

*CS Unplugged*

The book "Computer Science Unplugged...off-line activities and games for all ages", published in 1998 by the New Zealand computer scientists Bell, Witten and Fellows (1998), was the eponym and international pioneer for computer science lessons for children without computers. With the activities and games presented in the book, the authors mainly target primary school children aged 5 to 12 by explaining a wide range of computer science topics using everyday materials such as playing cards, string, pens and other household materials. Topics include algorithms, artificial intelligence, binary numbers, logic circuits, compression, cryptography, information representation and parallel processing.

As in demonstrations and hands-on experiments known from the natural sciences, the examples can help to demonstrate ideas of computer science in a stimulating and entertaining way in a "computer science show" for families (see Bell, 1999). In German-speaking countries, the ideas and materials of the CS-Unplugged approach have been picked up and partly advanced, e.g., in the Computer Science Year in the project "Einstieg Informatik" (Getting started in Computer Science), especially targeting children from 5 years of age (Pohl, Kranzdorf & Hein, 2007). Under the title "Abenteuer Informatik" (Adventure Computer Science), Gallenbacher developed a participatory exhibition on computer science where grade-3 children could play and understand (see Gallenbacher, 2009).

*Figure 13. An example of CS Unplugged: Sorting by weight using a beam scale: Since intuitive sorting, for example, by numbers, is not possible, complex comparison operations are explicitly carried out*

In computer science didactics worldwide, CS Unplugged has become one of the most popular and influential approaches for a playful introduction to computer science without a computer. It is applied in school and out-of-school learning situations, and is referred to internationally in computer science curricula and has been evaluated in various studies. For example, it was found that younger children in particular engage with the ideas and enjoy investigating them further (see Bell, Rosamond & Casey, 2012), that CS Unplugged can be used to spark an interest in computer science among fourth graders (Lambert & Guiffre, 2009), and that access even helps teachers to convey computer science ideas (Morreale & Joiner, 2011). It remains questionable, however, whether CS Unplugged will really achieve its long-term goal of motivating children to engage further with computer science and change the perception that computers are a tool, but not the centre of computer science (Taub, Ben-Ari & Armoni, 2009).

*Examples of computer science in materials for early-childhood education*
Scientific discussions around computer science education in early childhood education are still very young. Nevertheless, there are already various materials for early childhood and primary school education that deal with computer science topics. For example, the "Haus der kleinen Forscher" Foundation offers various activities that have a clear connection to computer science topics and are also partly derived from CS Unplugged (cf. Figure 14). On the one hand, these are materials that are thematically directly related to computer science (e.g., "Digitale Kommunikation"[18] (digital communication)) and, on the other hand, materials that describe facts and procedures that are important in other sciences as well as in computer science. For example, in the teaching and learning material on the topic "Mathematik entdecken: Modelle und Karten – Vom Gegenstand zum Symbol" (Discovering mathematics: models and maps – from object to symbol)[19], there are aspects on computer science modelling and representation of information. In the thematic brochure "Kommst du mit die Zeit entdecken" (come along to discover time) (HdkF, 2013b), important computer science topics, namely estimation effort, efficiency and optimisation are examined and the material on the topic of mathematics "Spannende Wiederholungen" (exciting repetitions)[20] deals, for example, with computer science ideas of algorithm and iteration.

Computer science-related tasks are also available in other learning materials for children, e.g., within the context of logical thinking. For example, the book of the same name in the "Kindergarten Lernraupe" (kindergarten learning caterpillar)-series for children from the age of 3 (Wiesner, 2008) is about describing processes (algorithmisation) and recognising the inputs of a processing procedure (IPO principle) (cf. Figure 15).

---

18 In the thematic brochure "Kannst Du mich verstehen? Die Vielfalt der Kommunikation erkunden und erforschen" (Can you understand me? Exploring and investigating the diversity of communication) (HdkF, 2014)
19 Cf. https://www.haus-der-kleinen-forscher.de/fileadmin/Redaktion/6_Experimente/Mathematik/Downloads/MATHE__KARTENSET_6.2014.pdf?pk_campaign=Newsletter%20August%202014&pk_kwd=Karten-Set-Mathe
20 Cf. http://www.haus-der-kleinen-forscher.de/uploads/tx_hdkfexp/110831_Spannende_Wiederholungen_Web.pdf

In this context, it should be noted that only inadequate research results are available in the area of such elementary skills or precursor skills (as they exist for mathematics, for example) that can be assigned to computer science.

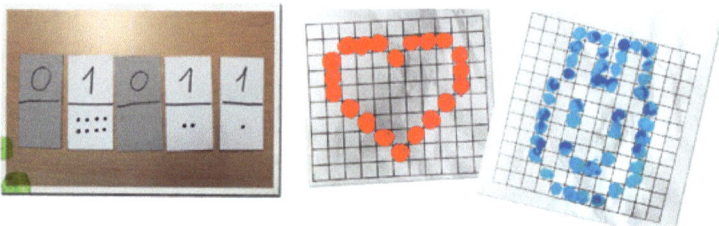

*Figure 14.* CS Unplugged tasks in the materials of "Haus der kleinen Forscher" Foundation: Binary numbers (left) and image representation with raster graphics (right) are shown on paper (Image source: HdkF, 2014)

*Figure 15.* Computer science-related tasks in "Lernraupe" (Learning Caterpillar) (Wiesner, 2008)

*Children's literature on computer science*
Only recently has computer science become an explicit topic in children's literature. Books on this subject are mainly available in English-speaking countries. "Hello Ruby: Adventures in Coding" (Liukas, 2015, see Figure 16), on the one hand, tries to convey computational thinking, while, on the other hand, it aims to teach the basics of the structure of informatics systems and programming, as well as terms and phenomena of the IT world at the child-care centre level. In addition to the age-appropriate story, the book contains various exercises and craft materials. "Lift-the-Flap Computers and Coding" (Dickins, Nielsen, Barden & Lamont, 2015; see Figure 17) is aimed at primary school children and explains the structure of hardware, as well as individual computer-related phenomena (programmes, binary systems, character coding etc.) in an age-appropriate way.

*Figure 16.* "Hello Ruby" teaches concepts of computer science with a special focus on "computational thinking" at the child-care centre level (Liukas, 2015, Image source: http://www.helloruby.com/press)

*Figure 17.* Flip-open book "Lift-the-Flap Computers and Coding"

*Bebras for children*

The annual international competition "Informatik-Biber"[21] (internationally "Bebras") has also been offered in Germany since 2016 for children from grade 3. Even though this is an online competition, i.e., a computer is used to present the tasks and to enter solutions, Bebras is mainly characterised by its puzzle tasks, most of which can also be carried out without a computer. In other countries, e.g., in Slovakia since 2010, the competition has already been used successfully in primary schools for several years (see Gujberova & Kalas, 2013). The competition features a collection of age-appropriate, partly interactive computer science tasks that must be solved within a few minutes and are intended to convey the diversity of computer science, establish the first contact and encourage further involvement with computer science (cf. Figure 18). The data collected during the

---

21  http://informatik-biber.de/

competition could provide interesting research results in the future about the cognitive performance of children when working on such computer science tasks.

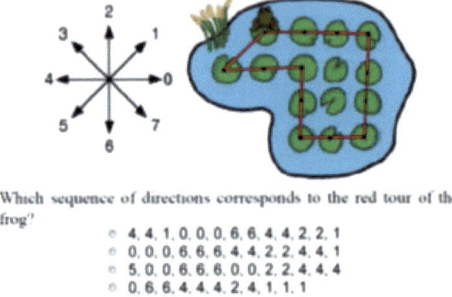

Figure 18. *Example of a task from Bebras for primary schools (Gujberova & Kalas, 2013)*

### 2.3.2 Software-Based Introductions to Programming

It is often tools that arouse children's curiosity and interest in computer science. Interactivity, direct feedback and extensive design options make informatics systems attractive learning media, which in many cases have also been developed for computer science teaching. In particular, there are numerous software tools that enable children to enter computer science through programming. Kelleher and Pausch analyse tools for learning programming and distinguish these with regard to two different goals associated with programming: "Teaching systems" help to learn programming as correctly as possible (Kelleher & Pausch, 2005). "Empowering systems" are primarily intended to support learners using programming as a creative tool. The tools described hereinafter belong to this second category. They are characterised by the fact that the complexity has been reduced in a meaningful way, e.g., by using visual building blocks that do not allow syntax errors.

*Background*
Computers were already used in learning contexts as early as the 1960s. Papert noted that these applications mainly fall into three areas: (1) Tutorials, in which the computer acts as an instructor, (2) software tools, such as calculators, text processing or simulations and (3) micro-worlds, which are snippets of reality that enable a new kind of learning, only made possible through computer technology (Papert, 1987). A large part of the software tools available for the introduction into computer science is based on the idea of micro-worlds, which Papert defined as a "...subset of reality or a constructed reality whose structure matches that of a given cognitive mechanism so as to provide an environment where the latter can operate effectively. The concept leads to the project of inventing micro-worlds so structured as to allow a human learner to exercise particular powerful ideas or

intellectual skills" (Papert, 1980, p. 240). Micro-worlds represent a manageable virtual learning space in which children can explore existing virtual objects and develop them further within the given possibilities. Programming is understood here as an intellectual tool for exploring and understanding often abstract issues.

With the programming language Logo (Feurzeig, Papert & Lawler, 1970), a micro-world for programming was created, for which there are extensive reports from the 80s on its use in primary schools (e.g., Hoppe & Löthe, 1984; Ziegenbalg, 1985). The aim of Logo was initially not so much to teach computer science, but rather an understanding of mathematical and geometric structures in the sense of the constructivist learning approach as an extension of the constructivist learning theory. Here, Papert emphasises the importance of active learning by using specific, personally meaningful objects: "[Learning] happens especially felicitously in a context where the learner is consciously engaged in constructing a public entity, whether it's a sand castle on the beach or a theory of the universe" (Papert & Harel, 1991, p. 1). In contrast to constructivism, it is emphasised that knowledge construction not only takes place in the learner's head but should be based on an actual construction process in the real or virtual world ("learning by making" instead of "learning by doing"). This advantage for the learner is that the resulting product can be tried out, shown, discussed, analysed and also admired, and the learner has the opportunity to become "one" with the observed phenomenon, instead of looking at it "from the outside". With the help of micro-worlds as a learning environment on computers, such "learning experiences can take place unimpeded of the complexities of the world" (Papert, 1998, p. 66). In this way, micro-worlds become "breeding worlds for knowledge" and can serve as a "greenhouse for a particular species of supporting ideas or intellectual structures" (Papert, 1982, p. 157).

Based on the "Logo" model, various other micro-worlds were developed in the 1980s and 1990s, some of which are still used today in computer science lessons, especially in the junior secondary level (e.g., Robot Karol, Kara, Java-Hamster; Boles, 2005; Reichert, Nievergelt & Hartmann, 2005). However, the one-dimensionality of many of these micro-worlds is problematic: It raises a questionable image of computer science if programming and informatics are primarily used to move ladybirds or robots through mazes to collect various objects. Likewise, some micro-worlds violate the underlying ideas: While in constructionist learning, learners create personally meaningful products by means of a micro-world, the possibilities in many micro-worlds are so limited that it is not a product that needs to be created but merely a solution that needs to be found for given, often artificial problems.

Many recently developed tools, such as the Scratch programming environment, overcome this shortcoming (see below). When used in lessons, these more recent tools have shown advantages, for example, in motivating learners (see Ruf, Mühling & Hubwieser, 2014).

*Scratch/ScratchJr*
A good example for a child-friendly introduction to programming is the Scratch[22] development environment (cf. Figure 19), which is based on Papert's ideas and mainly implements the following principles: Low Floors, High Ceilings, Wide Walls (see Resnick & Silverman, 2005), i.e., even complex, demanding projects should be realisable with the lowest possible starting hurdles (step by step) and different interests of children should be addressed. The low starting hurdle is achieved through a visual programming language that is largely intuitive and does not allow syntax errors – one of the biggest hurdle for programming beginners (see Myers, 1990). The programmes can be run at any time, so that learners can assess the success of their work themselves whenever they wish.

Compared to older micro-worlds, Scratch enables the implementation of a wide variety of multi-media projects: animations, storytelling, games, simulations and much more. In addition, projects can be shared on an online gallery and jointly edited. As such, the developers of Scratch were not so much interested in training programmers but rather in enabling children to deal creatively with digital media (see Resnick et al., 2009). The approach chosen in Scratch has proven to be very successful: Within a very short time, the programming environment was used worldwide.

Following this example, several other programming environments with a similar approach, but sometimes with a different focus, have been developed to help children learn programming, for example, by developing apps (e.g., with App Inventor[23]) or with online brain games (e.g., Blockly[24]). Since the graphic representation of the programming constructs can also offer advantages for more elaborate programming, various programming environments modelled on Scratch are currently being developed for continuing education and for occasional programmers (e.g., Snap[25], GP[26]). Since 2012, attempts have been made to reach children as young as 5 years old with this approach by developing an age-appropriate user interface with touch operation and exclusively iconic representation for which children do not need any reading skills (Flannery et al., 2013).

---

22 http://scratch.mit.edu
23 http://appinventor.mit.edu
24 https://blockly-games.appspot.com/
25 http://snap.berkeley.edu/
26 http://scratch-dach.info/wiki/GP_(Programmiersprache)

*Figure 19.* Visual programming using Scratch building blocks (left) and in Scratch Jr (right)

*Online tutorials*

With the aim of motivating budding programmers while providing the easiest possible access, several websites offer collections of instructed activities and learning units related to programming that are also of interest to schools (https://code.org/, http://start-coding.de/). These brief insights into programming became popular through the "Hour of code" campaign, which reached almost 100 million people worldwide by 2015 (Wilson, 2015).

### 2.3.3 Programmable Toys (Robotics)

McNerney reports on research conducted by Seymor Papert in the 1970s aimed at linking programming with the physical world in order to introduce children to the world of computer science (McNerney, 2004). Because primary school children were not yet really able to write programming instructions on a keyboard, the logo turtle was used prototypically as a drawing robot, and alternative physical ways of programming were tested, e.g., with buttons and plug-in cards (see Perlman, 1976). Since then, research has regularly focused on accessibility via physical experiences. It has proven to be an advantage to use that direct interaction with physical objects because it is easier and therefore less frustrating for children (Xie, Antle & Motamedi, 2008). So, it is not surprising that there are increasing attempts to develop "tangible programming interfaces", especially for younger children (Brauner, 2009; Leonhardt, 2015). For example, toys are now commercially available that can be programmed directly on the device, such as the BeeBot[27].

Successful advancements of the physical Logo turtle are robotic kits such as e.g., LEGO Mindstorms[28] or WeDo[29], from which further developments for programmable, interactive creative kits have recently emerged (e.g., Pico Crickets, see Resnick, 2007). What these tools have in common is that they provide a range of sensors (e.g., for volume, brightness, distance) whose data can be collected

---

27  https://www.b-bot.de/produkte/bee-bots/
28  https://www.lego.com/de-de/mindstorms
29  https://education.lego.com/de-de/products/lego-education-wedo-2-0-set/45300#wedo-20-set

and processed via the programming interface to trigger actions, e.g., to control motors, LEDs and speakers.

Against the background of the increasing disappearance of desktop computers and the omnipresence of computers as so-called embedded systems in everyday objects, such tools lend themselves to computer science education as exciting and contemporary, because children can use them to discover computer technology in everyday life and also design it themselves, e.g., as interactive toys. In the following, different examples of robotics kits and programmable toys are discussed in more detail.

### LEGO Mindstorms & WeDo

LEGO Mindstorms have been widely used in computer science classes to specifically implement the idea of construction (see Wiesner & Brinda, 2007). They are thus used to transfer the idea of micro-worlds from the virtual to the real world. What is criticised about the Mindstorms robots is that their construction is very complex and therefore difficult for younger children to understand. Based on their own experience, Borowski and Diethelm assume that this approach is not yet entirely suitable for 10-year-old children (Borowski & Diethelm, 2009). Even though very different constructions can be built with LEGO Mindstorms, often only robot vehicles are used (see Wiesner & Brinda, 2007). In Germany, LEGO Mindstorms are probably among the most widespread programmable robots and are used extensively in both school and non-school settings (Hartmann & Schecker, 2005; Leonhardt, 2015).

A simple version of the LEGO Mindstorms system is LEGO WeDo, recommended for children from the age of 7. So far, there are only a few reports on its use. Working with third graders, Mayerová observed that the children were basically able to solve the tasks set but sometimes spent more time putting the mechanical model together than programming it (Mayerová, 2012).

### PicoCrickets

PicoCrickets[30] can be seen as a new development, based on the experience with LEGO Mindstorms robots. PicoCrickets extend the idea of programmable building blocks to the world of omnipresent computers. PicoCrickets are small, programmable building blocks, to which sensors (light, touch, resistance, sound) can be connected as input sources. Children can use these bricks to design creations that move, light up, make music and much more. The simple visual programming language PicoBlocks is used to control its output options (coloured lights, motors, sound generator and LED display). This results in a variety of feasible projects, for

---

30 https://www.playfulinvention.com/picocricket/index.html (no longer produced).

example, interactive gardens, responsive soft toys, or "techno-clothes", such as boots described by Resnick that flash in colour depending on the walking speed (Resnick, 2007). PicoCrickets includes a set of LEGO building blocks and handicraft materials. This is to overcome the separation between the world of electronics and the world of handicrafts. It allows children to engage in a wide variety of creative and computer-related activities in contexts from their own world of experience. PicoCrickets is therefore not directly a form of robotics, but is presented as an "invention kit that integrates art":

> "You can plug lights, motors, and sensors into a Cricket, then write computer programmes to tell them how to react and behave. With Crickets, you can create musical sculptures, interactive jewellery, dancing creatures, and other artistic inventions – and learn important maths, science, and engineering ideas in the process" (MIT, 2011).

The authors themselves have collected the positive experiences with the use of Pico Crickets in the 4th grade of a primary school (Romeike & Reichert, 2011).

*Bee-Bot*
Bee-Bot is a mobile robot especially designed for young children and looks like a bee with its black and yellow stripes (see Figure 20). It is equipped with seven input buttons. It can be steered forwards, backwards, 90 degrees to the right or 90 degrees to the left. After entering a maximum of forty of these four commands in any combination, the Bee-Bot can be started by pressing the "Go" button. The two remaining keys are used to stop and reset the instructions. All Bee-Bot applications can be used to learn and practice the mental execution of commands. Since all commands must be entered before the "Go" key is pressed, a strategy is always worked out in advance and then converted into a sequence of commands.

*Figure 20.* Bee-Bot, the programmable floor robot

*Other robot systems: Cubetto, Dash & Dot*

The PRIMO robot Cubetto (see Figure 21 left) is also intended to ease the first steps, especially for young computer science beginners. It can be programmed by plugging in wooden blocks. With this simple educational toy, even young children can learn the basics of understanding algorithms (or programme sequences) and programming in a playful way.

*Figure 21. Different robotics systems for children (from left to right): PRIMO Cubetto, LEGO Mindstorms, WONDER robot Dash & Dot*

Especially for very young learners, direct feedback on their work promotes motivation (Leonhardt, 2015). Thus, even somewhat more complex robot models, such as the WONDER robots Dash & Dot (see Figure 21, right), which are controlled via symbols on a tablet, allow users to directly admire and evaluate the result of their own programming attempts.

### 2.3.4 Out-of-School Learning Venues and Communities

While the approaches described above are oriented towards specific examples and topics (CS Unplugged) or tools, other approaches focus on social and collaborative experiences to motivate children and young people for computer science, especially in extracurricular settings. Here, different types of online and offline communities offer children and young people the opportunity to exchange ideas with other interested people and to have collaborative learning experiences.

*Computer Clubhouses*

One example of this is *Computer Clubhouses*[31], which also allow children at an early age to participate in offers to learn with computers. The project "Computer Clubhouses" aims to enable learners to learn creatively with computers (Resnick & Rusk, 1996). The Clubhouse learning approach is based on four guiding principles: Learning by design, pursuing one's own interests, building a community

---

31 https://theclubhousenetwork.org/

and an environment of respect and trust. Even though computer science education is not mentioned as the primary goal of the Clubhouse learning approach, but rather intercultural understanding and tolerant coexistence are aimed at, the participants learn basic computer science procedures, concepts and applications virtually "along the way".

*Learner laboratories*
Another growing institution in the field of computer science are *learner labs*, such as the *InfoSphere*[32] learner lab at RWTH Aachen University, which solely aims at teaching aspects of computer science to children and adolescents from grade 3 upwards (Bergner, 2015). Learner labs often offer workshops to entire school classes but also for individual interested learners.

*Competitions*
In addition to the "Informatik Biber" (computer-science Bebras) competition described in 2.3.1, there are various other offers. For example, the Federal Computer Science Competition[33] challenges advanced computer science learners. There are also many regional and international robotics competitions, some of which are offered directly by companies (e.g., LEGO), but also by (variously sized and professional) groups, networks or even schools themselves.

*Online Communities*
Brennan noted that apart from the process of creating Scratch projects being an important reason for children to get involved with programming, the opportunity the platform offers to connect with others and share ideas is also important (Brennan, 2013). In the meantime, the Scratch community has hundreds of thousands of members, mostly aged 8 to 16, who upload more than 2,000 projects to the website every day, where they can be commented on, praised or further edited (Brennan, 2013). The source code of the projects can be viewed by anyone, so that one can learn from the others. Kafai and Burke emphasise that, according to their observations, it is the social experience rather than the tools that are important for learning programming (Kafai & Burke, 2014). Programming increasingly presents itself as a very social rather than an individual endeavour. The exchange and especially the positive feedback apparently strengthens the positive appreciation of one's own creative process, similar to presentations in front of an audience, and gives children a new self-competence in dealing with technology.

---

32  http://schuelerlabor.informatik.rwth-aachen.de/
33  http://www.bwinf.de/

*Maker Culture*

The so-called maker culture is about enabling people of all ages to invent and create things, especially with new technologies. Typically, this is not done in school but in study groups, public institutions, (such as e.g., FabLabs; Walter-Herrmann & Büching, 2013), at fairs and exhibitions (Maker Faire[34]) or privately. Since computer science-based technologies, such as 3D printers, micro-controller boards or robotics resonate strongly in the Maker Community, there is a lot of overlap with examples from computer science education (see Libow Martinez & Stager, 2013).

### 2.3.5 Summary

Even if computer science education for children at the level of child-care centre and primary school is a relatively new field of research, there are already various tried and tested approaches that can motivate children and teach them computer science skills. With regard to the suitability of the different target groups, there are approaches that are intended and tested for children as early as at the child-care centre level (e.g., learning materials of early childhood education, "Hello Ruby", Bee-Bot, ScratchJr), as well as offers which are mainly aimed at children of primary school age (often around grade 3), e.g., CS Unplugged, "Informatik Biber" (Bebras), LEGO WeDo, Scratch, learner labs). However, the research situation in this area can be described as insufficient because the various field reports on practical use can at best provide initial indications of the design and impact of computer science education in this age group.

## 2.4 International Comparison: Curricula and their Classification in the Competence Model

In contrast to mathematics and natural sciences, there is no long-established universal compulsory subject of computer science in the junior secondary level – neither nationally, nor internationally. As a result, there is much less experience and study on which computer science skills can be acquired at what ages and how this can be achieved. Consequently, there is little firm knowledge about which computer science competencies or competencies fundamental to computer science can already be acquired at the primary level or even at pre-school age (cf. Section 2.2).

Nevertheless, as described in the previous sections, there are studies that show that children can learn concepts of computer science and that it makes sense to do so at a young age, and there are specific examples and approaches to teach these concepts in an age-appropriate way.

---

*34 For Germany: https://maker-faire.de/*

Based on this, some progressive education systems at the international level are currently introducing computer science education in junior secondary and primary education as a general compulsory subject to teach learners a computational way of thinking and problem-solving skills using computer science methods (internationally referred to as "computational thinking"). This way of thinking  qualifies the learners both for well-founded and reflective handling of informatics systems and for the use of informatics systems to creatively shape their own life-world. For this reason, the subjects are generally referred to by terms such as Computer Science, Informatics, Programming, Computing (UK, NZ), Computational Thinking (USA) or Digital Technologies (AUS). Regardless of their designation, the international initiatives for computer science education include, on the one hand, operating and media competencies (digital literacy and basic ICT education), and, on the other hand, problem-solving competencies and technical design competencies (computing, computational thinking), each of which must be reflected in the context of use (cf. Dagstuhl Triangle, Figure 12 from Section 1.6).

Most of the well-known approaches that are now included in international curricula come from extracurricular activities (club houses, summer camps, learning labs, coding initiatives, competitions, online communities, programmable toys). Despite such measures, which individually have been demonstrably successful (cf. Section 2.3), it has not been possible to trigger a stable interest in computer science in the long term. In contrast, this has been achieved in recent years in the other STEM disciplines, which are also supported by extracurricular measures. Mathematics and physics are more strongly opted for, while computer science continues to decline (Brown, Sentance, Crick & Humphreys, 2014, p. 3). The obvious difference between these disciplines is that computer science is not a regular school subject at the primary and junior secondary level.

Some of the international principles and standards, as well as curricula mentioned are analysed below to identify internationally recognised core content and competencies for the expert opinion. They are then presented in the outline of the GI's (GI – Gesellschaft für Informatik e.V., 2008) competence model for computer science education at the junior secondary level. From this analysis and representation, it can be deduced whether and how the competence model for the secondary level can be mapped onto one for the primary level. The aim is to define a competence structure for the primary level that is compatible with the established GI model for the junior secondary level, for which a preliminary education in computer science in the level of child-care centre and primary school should qualify.

### 2.4.1 Computing in the UK

The CAS[35] (Computing at School) initiative was founded in 2008 by BCS (British Chartered Institute of IT; comparable to the German Gesellschaft für Informatik) and IT companies Microsoft, Google and Intellect, with the intention of bringing computer science into schools. Previously, ICT in British schools was more in the sense of training the use of informatics systems. Just like in Germany, ICT curricula were not very successful (e.g., recommendations of the The Royal Society, 2012). They tended to discourage children and young people from taking an interest in computer science and its methods, and gave a false image of the discipline.

The CAS initiative initially consisted of a few committed teachers and academics, but quickly grew to hundreds and today has over 6,000 members. In the beginning, the CAS consortium exerted political influence, mainly by making it clear that computer science is a scientific discipline that develops its own ideas, principles, techniques and methods that are applied in wide areas of life. It is based on its own kind of computational thinking, which can generally be used to solve problems (even without the use of informatics systems) and is increasingly finding its way into many other disciplines.

This led to the questioning of previous ICT teaching (operating competence for informatics systems) and the development of new national curricula for the subject "computer science" from primary school onwards, as well as the introduction of the subject at primary schools from 2014. In these curricula, important user competencies (ICT and digital literacy) were retained as essential foundations but expanded to include computer science competencies, so that informatics systems and the computer science concepts realised in them can also be understood and used reflexively to creatively design solutions for upcoming tasks.

In order to illustrate the competencies of computer science education, the topics of the field (key concepts: *languages, machines, and computation, data and representation; communication and coordination; abstraction and design; computers and computing are part of a wider context*) and typical actions and ways of thinking (key processes: *abstraction, modelling, decomposing, generalising, and programming*) are described. From this, the content and scope for a pre-education in primary and secondary education are derived, which include the topics

---

[35] http://www.computingatschool.org.uk/

of algorithms, programmes, data, computers, communication and the Internet, as well as "in-depth topics".

For the educational levels to be achieved, the CAS concept distinguishes nine target competence levels (level 1-8 plus exceptional) for the age groups (Key Stages 1-4) (depending on the hours available in the school) from the primary level onwards:

- Key Stage 1: In pre-school to grade 2 (age 5-7), competence domains 1 to 3, if applicable, can be achieved
    - Level 1: Discuss sequences (storyboards), arrange objects, recognise objects that process input, use programmable toys
    - Level 2: Draw own sequences (storyboards), give direct instructions, use programmable toys to perform specific tasks, classify objects

- Key Stage 2: In grades 3-6 (age 7-11), the aim is to achieve competence levels 2 to 4 accordingly
    - Level 3: Identify commonalities in storyboards, plan a series of instructions, give linear instructions, present data systematically
    - Level 4: Analyse and represent a sequence of events, recognise different types of data (text, number), understand the precision of a programming language syntax, give instructions with selection and repetition, think through an algorithm and predict the result

Building on this, the further levels of competence are to be achieved in continuous computer science lessons in secondary school:

- Key stage 3: In grades 7-9 (age 11-14) competence level 5 to 6
    - Level 5: Break down problems and present them in appropriate notation, analyse and explain selected algorithms, recognise similar problems and identify algorithms that can be used for them, explore variables in a programme, develop and test sequences of instructions step by step
    - Level 6: Describe slightly more complex problems (searching and sorting), graph system components, specify models for similar problems, analyse programmes and predict behaviour, compare different solutions, use parameterised procedures and functions

- Key Stage 4: In grades 10-12 (ages 14-17), corresponding to our senior secondary level, further competency levels 7 to 8 and, for exceptionally talented learners in more in-depth computer science courses, the *exceptional* level

- Level 7: Describe search and sort algorithms, break down problems using appropriate notation, recognise similarities in slightly more complex problems, assemble given programme modules, use more complex data structures including relational databases, select programming tools appropriately, text-based programming
- Level 8: Select appropriate programming constructs, find suitable models for more complex problems, advanced troubleshooting and debugging, analyse and optimise more complex data structures, understand the relationship between reality, the model, logic, algorithm and visualisation
- Exceptional: use professional programming language, specify general models for problem categories

The English Ministry of Education has taken up the Commission's proposals and introduced an end-to-end *computing* subject from 2014, which identifies the following competencies as mandatory in the areas relevant to the expert report (Key Stages 1 and 2) (Department for Education, 2013):

- Key Stage 1: Pre-school to grade 2 (age 5-7)
    - Understand what algorithms are and how they are implemented in the form of programmes on digital computers (devices). Understand that programmes follow precise and unambiguous instructions.
    - Develop and test own simple programmes.
    - Draw logic conclusions to predict the behaviour of simple programmes.
    - Use digital technologies purposefully to create, organise, store, access and adapt digital content.
    - Be aware of the use of digital technologies in everyday life.
    - Use digital technologies safely and respectfully, keep data confidential and know who to contact if there are concerns about content or contact requests via the Internet.

- Key Stage 2: In grade 3-6 (age 7-11)
    - Design, develop, implement and test programmes that fulfil specific goals, including simulations of physical systems; solve problems by breaking them down.
    - Use sequences, branches and loops in programmes; develop programmes with variables and different forms of input and output.
    - Explain the behaviour of simple algorithms through logical reasoning, find errors in algorithms and programmes and correct them.

- Understand computer networks, including the Internet, how they provide various services, such as the WWW and assess their potential for communication and collaboration.
- Use search technologies effectively and assess the selection and ranking of search results; evaluate digital content critically.
- Select, combine and use software tools (including web services) on different informatics systems (PC, tablet, smartphone), to create and manage a range of digital objects (programmes, systems, content) purposefully. This includes collecting, analysing, evaluating and presenting information and data.
- Use digital technologies safely, respectfully and responsibly; recognise acceptable and unacceptable behaviour and know how to report concerns.

A number of teaching ideas and materials have been developed for the curriculum[36] and corresponding advanced training for teachers has been designed. All content points always concern aspects of reflective use – digital literacy (DL) and informatics technology (ICT), as well as basic concepts of computer science (CS). In the beginning, the focus is more on user competencies and, with increasing age and competence level, also on concepts of computer science (see Table 1). For the age range considered in this expert report, competence levels 1 to 3 (and sometimes 4) are particularly relevant.

---

36  http://community.computingatschool.org.uk/resources

|   | CS | ICT | DL |
|---|---|---|---|
| 1 | Understand what algorithms are Create simple programs | Use technology purposefully to create digital content<br>Use technology purposefully to store digital content<br>Use technology purposefully to retrieve digital content | Use technology safely<br>Keep personal information private<br>Recognise common uses of information technology beyond school |
| 2 | Understand that algorithms are implemented as programs on digital devices<br>Understand that programs execute by following precise and unambiguous instructions<br>Debug simple programs<br>Use logical reasoning to predict the behaviour of simple programs | Use technology purposefully to organise digital content<br>Use technology purposefully to manipulate digital content | Use technology respect fully<br>Identify where to go for help and support when they have concerns about content or contact on the internet or other online technologies |
| 3 | Write programs that accomplish specific goals<br>Use sequence in programs<br>Work with various forms of input<br>Work with various forms of output | Use search technologies effectively<br>Use a variety of software to accomplish given goals<br>Collect information<br>Design and create content<br>Present information | Use technology responsibly<br>Identify a range of ways to report concerns about contact |
| 4 | Design programs that accomplish specific goals<br>Design and create programs<br>Debug programs that accomplish specific goals<br>Use repetition in programs<br>Control or simulate physical systems<br>Use logical reasoning to detect and correct errors in programs<br>Understand how computer networks can provide multiple services, such as the World Wide Web<br>Appreciate how search results are selected | Select a variety of software to accomplish given goals<br>Select, use and combine internet services<br>Analyse information<br>Evaluate information<br>Collect data<br>Present data | Understand the opportunities computer networks offer for communication<br>Identify a range of ways to report concerns about content<br>Recognise acceptable/unacceptable behaviour |

| CS | ICT | DL |
|---|---|---|
| 5 Solve problems by decomposing them into smaller parts<br>Use selection in programs<br>Work with variables<br>Use logical reasoning to explain how some simple algorithms work<br>Use logical reasoning to detect and correct errors in algorithms<br>Understand computer networks, including the internet<br>Appreciate how search results are ranked | Combine a variety of software to accomplish given goals<br>Select, use and combine software on a range of digital devices<br>Analyse data<br>Evaluate data<br>Design and create systems | Understand the opportunities computer networks offer for collaboration<br>Be discerning in evaluating digital |

*Table 1. Competence levels 1-5 are broken down into computer skills (CS), reflective use (DL), and basic knowledge of information technologies (ICT) (Computing at School Working Group, 2013, p. 25).*

### 2.4.2 Computational Thinking in the USA

In the United States, too, there is discussion about whether and how a universal subject of computer science can be introduced, possibly even starting at the primary level. The efforts of associations and teachers' unions are receiving prominent support: Former President Obama took part in an *"Hour of Code"* and, according to the White House website, became the first American president to programme himself. In the process, he announced a 4-billion-dollar programme to introduce computer science in schools over the next few years[37]:

> *"In the coming years, we should build on that progress, by … offering every student the hands-on computer science and maths classes that make them job-ready on day one. … give all students across the country the chance to learn computer science (CS) in school … recognizing that CS is a 'new basic' skill necessary for economic opportunity and social mobility."*

The current situation regarding computer science education in general education schools in the USA is generally considered to be problematic and inadequate (Wilson, Sudal, Stephenson & Stehlik, 2010). As in Germany, there are differences be-

---

[37] https://www.whitehouse.gov/the-press-office/2016/01/30/fact-sheet-president-obama-announces-computer-science-all-initiative-0

tween the individual states. In the report "*Running on Empty*", it becomes evident that in more than a third of the states, neither ICT nor computer science education is offered in schools, in another third, there is only basic ICT education and in the remaining third, computer sciences is partially offered as an elective subject. Because of this situation, computer science in the US has lost young talent in recent years.

The ACM (Association for Computing Machinery, comparable to the German GI) offers a curriculum for secondary schools and advanced placement exams. The Computer Science Teachers Association (CSTA) has developed recommendations for computer science standards to serve as a framework for the states' future curriculum design ("CSTA K-12 Computer Science Standards", 2011). The proposal aims at a universal K-12 curriculum. The document criticises the untenable state of lack of computer science education in the US. Most states currently teach only user skills (digital literacy and ICT), while computer science competencies is lacking. There is often a misconception among politicians, parents and teachers that these disciplines are no different at all: .... *"general public is not as well educated about computer science as it should be, to the point that the nation faces a serious shortage of computer scientists at all levels that is likely to continue into the foreseeable future"*. Just as with the British proposals presented earlier, the predominant offerings, if any, in curricula for pure IT use skills are enriched by the focus on computer science education. The American and British working groups exchanged ideas and enriched each other's work.

The CSTA standards address the fields of (1) *Collaboration*, (2) *Computational Thinking*, (3) *Computing Practice and Programming* and (4) *Community, Global and Ethical Impacts*. It is emphasised that children love computer science education (computing) because of the versatile, often creative facets: *"... the combination of art, narrative, design, programming, and sheer enjoyment that comes from creating their own virtual worlds"*.

The recommendations for computer science standards distinguish between three competence levels, which in turn are divided into six age segments. Of these, the first two (level 1:3 for the age group 5-8 years (K-3) and level 1:6 for the age group 8-11 years (grades 3-6)) are relevant for the age range considered in this expert report and the level of competence to be achieved. In the first level of competence, called "Computer Science and Me" (K-6; pre-school to grade 6, for 5-11-year-olds), the basic concepts of computer science should be introduced to all learners from the primary level onwards. The expert group for CSTA recommendations assumes that this can be achieved by teaching the use and application of technology and the underlying concepts of computer science by integrating this into other school subjects.

In the first age level (K-3), comparable to Key Stage 1 of the UK curriculum, the four competence domains are described:

- The first age level of the *Collaboration* competence domain is about using programmes to gather information and to communicate and collaborate electronically.
- The *Computational Thinking* competence domain involves solving age-appropriate logical tasks using programmes and apps, arranging and managing information, using programmes to put one's own stories into pictures and texts and understanding that computers process information in the form of 0s and 1s and that programmes are developed to control computers (including smartphones and tablets).
- *Computing Practice and Programming* includes competencies that relate more to the use of tools to create, arrange and manage information and digital representations and explicitly to their use in learning (more ICT/Digital Literacy in our classification).
- The competence level Community, Global, and Ethical Impacts is also more of an ICT literacy and describes the recognition of ethically correct use of IT tools (including the Internet).

In the age group up to grade 6 (comparable to Key Stage 2 of the UK curriculum), the following competencies are targeted:

- In the Collaboration domain, standard tools, such as word processing, presentation software and spreadsheets, including access to online resources, should be used individually and collaboratively.
- The Computational Thinking competence domain is most closely related to computer science education. At this age level, it involves understanding how algorithms are constructed from individual steps, breaking down problems and the use of simulations to solve problems.
- Again, the focus is on the competence domain Computing Practice and Programming, which essentially involves the proficient assembly and use of

computer science tools to create, manipulate and manage digital objects and their relevance to one's learning and application at work. Programming is described in terms of visual languages in only one of the nine items.

■ The competence domain Community, Global, and Ethical Impacts extends the basic competencies of the preliminary level to include discussion of ethical use, as well as critical reflection on the impact and security of informatics systems.

Examples on all topics are available at CSTA Web Repository of teaching materials: https://www.codes-isss.org/csta_subdomain/WebRepository/WebRepository/.

In comparison to the UK curriculum, it is generally noticeable that the CSTA recommendations in both age groups place more emphasis on the use of tools (*Digital Literacy & ICT*) and that significantly fewer computer skills are explicitly described (programming, programme constructs, algorithms, predicting the behaviour of programmes). Instead, more emphasis is placed on the use of information technology for collaboration (explicit topic of *Collaboration*) and the recognition of information technology tools in everyday life and professions, where computer science education is helpful. Furthermore, CSTA explicitly mentions binary representation. The latter three domains are not found in the UK proposals.

## 2 Foundation of Goals on the Children's Level

| | Level 1:3 K-3 (Age 5-8) | Level 1:6 3-6 (Age 8-11) |
|---|---|---|
| Collaboration | Gather information and communicate electronically with others with support from teachers, family members, or student partners. | Use productivity technology tools (e.g., word processing, spreadsheet, presentation software) for individual and collaborative writing, communication, and publishing activities. |
| | Work cooperatively and collaboratively with peers, teachers, and others using technology. | Use online resources (e.g., email, online discussions, collaborative web environments) to participate in collaborative problem-solving activities for the purpose of developing solutions or products. |
| | | Identify ways that teamwork and collaboration can support problem solving and innovation. |
| Computational Thinking | Use technology resources (e.g., puzzles, logical thinking programs) to solve age-appropriate problems. | Understand and use the basic steps in algorithmic problem solving (e.g., problem statement and exploration, examination of sample instances, design, implementation, and testing). |
| | Use writing tools, digital cameras, and drawing tools to illustrate thoughts, ideas, and stories in a step-by-step manner. | Develop a simple understanding of an algorithm (e.g., search, sequence of events, or sorting) using computer-free exercises. |
| | Understand how to arrange (sort) information into useful order, such as sorting students by birth date, without using a computer. | Demonstrate how a string of bits can be used to represent alphanumeric information. |
| | Recognize that software is created to control computer operations. | Describe how a simulation can be used to solve a problem. |
| | Demonstrate how 0s and 1s can be used to represent information. | Make a list of sub-problems to consider while addressing a larger problem. |
| | | Understand the connections between computer science and other fields. |

| | Level 1:3 K-3 (Age 5-8) | Level 1:6 3-6 (Age 8-11) |
|---|---|---|
| Computing Practice, Programming | Use technology resources to conduct age-appropriate Research. | Use technology resources (e.g., calculators, data collection probes, mobile devices, videos, educational software, and web tools) for problem-solving and self-directed learning. |
| | Use developmentally appropriate multimedia resources (e.g., interactive books and educational software) to support learning across the curriculum. | Use general-purpose productivity tools and peripherals to support personal productivity, remediate skill deficits, and facilitate learning. |
| | Create developmentally appropriate multimedia products with support from teachers, family members, or student partners. | Use technology tools (e.g., multimedia and text authoring, presentation, web tools, digital cameras, and scanners) for individual and collaborative writing, communication, and publishing activities. |
| | Construct a set of statements to be acted out to accomplish a simple task (e.g., turtle instructions). | Gather and manipulate data using a variety of digital tools. |
| | Identify jobs that use computing and technology. | Construct a program as a set of step-by-step instructions to be acted out (e.g., make a peanut butter and jelly sandwich activity). |
| | Gather and organize information using concept-mapping tools. | Implement problem solutions using a block-based visual programming language. |
| | | Use computing devices to access remote information, communicate with others in support of direct and independent learning, and pursue personal interests. |
| | | Navigate between webpages using hyperlinks and conduct simple searches using search engines. |
| | | Identify a wide range of jobs that require knowledge or use of computing. |

|  | Level 1:3 K-3 (Age 5-8) | Level 1:6 3-6 (Age 8-11) |
|---|---|---|
| Community, Global, and Ethical Impacts | Practice responsible digital citizenship (legal and ethical behaviors) in the use of technology systems and software. | Discuss basic issues related to responsible use of technology and information, and the consequences of inappropriate use. |
|  | Identify positive and negative social and ethical behaviors for using technology. | Identify the impact of technology (e.g., social networking, cyber bullying, mobile computing and communication, web technologies, cyber security, and virtualization) on personal life and society. |
|  |  | Evaluate the accuracy, relevance, appropriateness, comprehensiveness, and biases that occur in electronic information sources. |
|  |  | Construct a program as a set of step-by-step instructions to be acted out (e.g., make a peanut butter and jelly sandwich activity). |

*Table 2. The CSTA competence levels and domains included in this expert report*

### 2.4.3 Digital Technologies in New Zealand (& Australia)

Concurrently with the preparation of this report, curricula for Australia (http://www.australiancurriculum.edu.au/technologies/digital-technologies/curriculum/f-10) and New Zealand were also developed and proposed for implementation; Duncan & Bell, 2015). They are based on the USA and UK initiatives described above but structure the competence domains slightly differently to facilitate the connection to existing models of secondary education competence level in the domains of digital technologies. In both drafts, competencies in the use of computer tools and digital technologies are described in addition to the purely computer science aspects. In the first level, the Australian curriculum places a stronger emphasis on aspects of how digital technologies work (binary representation), similar to the CSTA proposal for the USA.

In the New Zealand draft, content is divided into six categories: (1) *Algorithms*, (2) *Programming*, (3) *Data Representation*, (4) *Digital Device Infrastructure*, (5) *Digital Applications*, and (6) *Humans and Computers*. These categories are intended to reflect the core principles of computer science: "*This classification corresponds to the key ideas in computation, since digital devices apply algorithms to data through the practical means of programming, and they produce digital content which must then be considered in the context of its impact on the individual and society*" (Duncan & Bell, 2015).

The proposed model targets the existing standards that describe computer science competencies for ages 15 and older. The targeted competencies are differentiated into six categories for five underlying *levels*, with the proposal distinguishing between a more conservative and a more progressive, advanced approach. For the purposes of this report, the first two levels of competency are again of interest.

On NCZ level 1 (age 5-7)

1. Understand what *Algorithms* are and be able to follow (interpret) them.

2. In the domain of *Programming*, initial small programmes for cybernetic systems (robot, turtle) should be developed. In the "advanced proposal", simple iteration should already take place and debugging is to be dealt with.

3. The domain of *Data Representation* is not yet considered in this age group in the conservative proposal; in the advanced proposal, 0s and 1s are to be understood as representations of texts and images.

4. *Digital Device Infrastructure* describes the (physical, motor) skills for operating digital systems (gestures, clicking, mouse, keyboard and touch screen). In the advanced proposal, the software and hardware components should be described additionally.

5. *Digital Applications* comprises competencies to create, organise, modify and access digital content. In the advanced proposal, multimedia is explicitly mentioned as digital content.

6. *Humans and Computers* primarily aims to enable the safe use of information technology by protecting personal information and raising awareness of misuse fears. The advanced proposal additionally includes recognition of use in everyday life.

At the following NCZ level 2 (ages 7-9), these domains of competence should be extended according to age:

1. Problems should be broken down and errors in algorithms found and corrected. In the advanced proposals, steps for problem-solving should be considered additionally.

2. *Programming* in visual environments should be extended to variables, iteration and case discrimination. In the less conservative approach, interactive programmes should be developed and tested.

3. It should be understood how binary representation is used as *Data Representation*. In the advanced proposal, this should be done for different forms of information (text, image, symbols).

4. The operation of *Digital Device Infrastructure* should be extended to include data acquisition and data transmission and in the advanced case, contain problem search and resolution.

5. Building on previous experience, *Digital Applications* should be combined to include search and, in the less conservative approach, advanced search and the presentation of results in simple tables and charts.

6. The domain *Humans and Computers* should be expanded to include respectful and responsible interaction. In the advanced proposal, this includes social needs analysis and information sharing.

The New Zealand model is largely consistent with the American and British models. Compared to the CSTA proposal, just as in the British proposal, the aspect of presenting information in the form of digital data is not addressed in the first age level. In return, similar to the American proposal, more attention is paid to the domain of operating digital technologies. The "less conservative" proposal goes a little further in the individual domains than the American one.

The Australian model distinguishes between "understanding of & knowledge about" and "skills and abilities in using" digital technologies. The age levels are set in the same way as in the New Zealand model: F-2 (ages 5-7) and Year 3 and 4 (ages 7-9). All in all, this proposal, which was implemented in 2016, is largely in line with the other models.

In level F-2, learners should gain experience in creating, managing and using digital objects through the playful use of informatics systems (including robots, programmable toys, etc.), as well as through collecting and organising data and displaying it in multimedia. In addition, they should learn the steps to algorithmic problem-solving and be able to describe the use of informatics systems in everyday life. Ethically responsible, and above all, safe handling of digital data is to be practised when it is used for communication and learning.

At the end of the next level (Years 3-4), learners are to be able to access the facets of informatics systems (hardware and software) for their use and explain how information can be represented in the form of data in a targeted manner. They should be able to develop and implement simple (partly interactive) programmes for simple problems and explain design decisions. They can use computer science tools safely and purposefully.

### 2.4.4 Swiss Curriculum 21 ("Lehrplan 21")

In the Curriculum 21 ("Lehrplan 21") project, the German-Swiss Conference of Cantonal Ministers of Education (D-EDK) developed Curriculum 21 from 2010 to 2014. Initially, it did not include any computer science education for the early levels. After protests by some associations and studies by the Hasler Foundation, a commission then expanded Curriculum 21 to include a module on "Media and Computer Science" from grade 3. The module contains application competencies (rather ICT use and media education) and basic competencies (rather computer science background).

The *ICT Literacy* domain is already addressed as a prerequisite of experience and competencies in computer science in cycle 1 (grades 1-2), and the following competencies are described:

- turning devices on and off, starting, operating and closing programmes and using simple functions
- logging on and off
- saving and retrieving documents
- dealing with simple control elements (GUI)

For the domain of computer science education from cycle 2 (grades 3-4) onwards, the following competencies are essentially described for the domains of (1) Data, (2) Algorithms and (3) Computer Systems: Learners can

- represent, structure and evaluate data from their environment
  - in different forms of representation,
  - distinguish between corresponding file formats
  - and know tree (directories) and network (web) structures.

- analyse simple problems, describe possible solution procedures and implement (and test) them in programmes.
  - understand processes with branches and loops
  - interpret algorithms and parameters
  - understand programmes and unambiguous instructions

- understand the structure and functioning of concepts of information processing and apply concepts of secure data processing.
  - distinguish between operating system level and application level

- understand different types of memory (e.g., hard disks, flash memory, main memory) and their advantages, disadvantages and size limitations
- apply strategies for solving problems with devices and programmes (e.g., help function, research)
- explain how data can be lost and know the most important measures to protect against this

Compared to the approaches listed above, the Swiss proposal is somewhat shorter in terms of computer science but is still broader than the status quo in Germany. In the Swiss draft, it is clearer than in the other international proposals that digital education is based on three different, but not sharply delimitable, domains of competence: (digital) media education, ICT application competence and computer science education. This view is also represented in the Dagstuhl representation of the GI and in other strategy papers of the GI Department of Computer Science and Education (cf. Section 1.6).

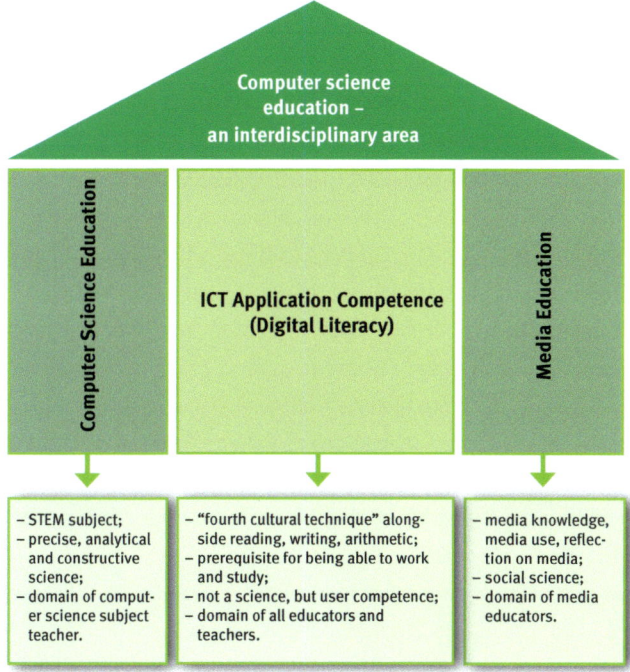

*Figure 22.* Computer science education as an interdisciplinary area (from Deutschschweizer Erziehungsdirektion, 2015)

## 2.5 Placing the International Standards Within the Framework of a Competence Model for Computer Science Education at the Primary Level

In the following, we will first present the competency structure model that underlies the recommendations for educational standards in computer science at the junior secondary level (GI – Gesellschaft für Informatik e.V., 2008), published in 2008. These recommendations currently form the essential basis for all drafts of the federal states of Germany (Bundesländer) that provide for computer science at the junior secondary level. However, it should be critically noted that the GI recommendations assume a continuous subject of computer science at the junior secondary level, whereas in most federal states computer science exists only as an optional subject in grades 8 and 9. In addition, many schools have ICT application courses in grades 5 and 6 either as independent subjects or integrated into others. Therefore, the competencies formulated in the GI recommendations are in fact not achievable. When designing computer science lessons at the junior secondary level, a selection is therefore usually made from the overall catalogue.

### 2.5.1 Competence Structure Model for Computer Science Education at the Secondary Level

The recommendations for computer science standards are based on the NCTM competence structure model (NCTM – National Council of Teachers of Mathematics, 1989, 1991, 1995, 2000) of US mathematics, which also forms the basis of the educational standards for mathematics in Germany. The NCTM and GI recommendations each distinguish between five Content and Process Domains. "The Content Domains characterise at least the subject competencies to be acquired. The Process Domains describe the way in which learners are to deal with the subject content mentioned" (GI – Gesellschaft für Informatik e.V., 2008). The competencies to be acquired can thus be described as a combination of Content and Process Domains (see Figure 23). Here, the respective domains cannot be clearly distinguished from each other.

The competencies are described for grades 5-7 and 8-10. The overall aim is to prepare all learners for life in a digitally shaped world. In doing so, they should understand the basic structure and functioning of informatics systems in order to

> "enable their targeted use for solving problems on the one hand, but also easy access to other systems of the same application on the other hand. However, the school-based examination of the structure and functioning of informatics systems must not only take place on the level of the

user interface, which can already change with the next product version or when using a product from another manufacturer" (GI – Gesellschaft für Informatik e.V., 2008, p. 19).

The starting point for a general education approach is the representation of "information" on questions from the learners' life world through "data" (Content Domain 1 – C1). Furthermore, general education includes the realisation that data and algorithms (C2) must be formulated in a specific, precisely defined language (C3) in order to be automatically processed by a computer or a simple model, such as that of automata (C3). Informatics systems of various Application Domains (C4) can be modelled and simple parts implemented (Process Domain 1 – P1), the use of which essentially shapes coexistence in our society (C5). In doing so, opportunities and risks are to be weighed up and possible design decisions are to be justified and evaluated (P2). These Content Domains should be addressed in lessons that encourage learners to "communicate" (P4) using appropriate computer science terminology, "structure" (P3), "reason", "evaluate" and "cooperate", as well as to "link" internal computer science knowledge with external knowledge.

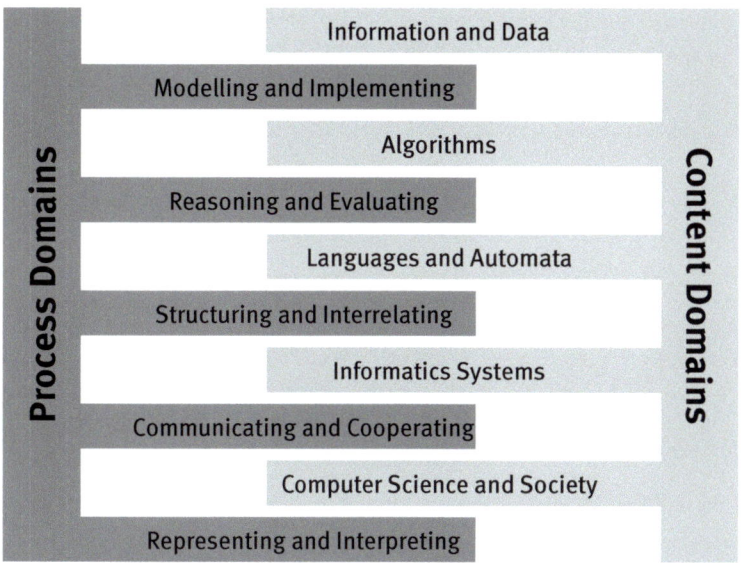

*Figure 23.* Recommendations for educational standards in computer science at the junior secondary level from www.informatikstandards.de

The structure was developed in accordance with the NCTM standards in several workshops with teachers and subject didactics experts over about five years. The domains of competence are defined individually across all grades in the following recommendations (GI – Gesellschaft für Informatik e.V., 2008, p. 20). The five Content Domains describe what the learners engage with:

*Information and Data (C1)*

Learners of all grades

- understand the relationship between information and data, as well as different ways of representing data,
- understand operations with data and interpret them in relation to the information presented,
- perform operations on data appropriately.

*Algorithms (C2)*

Learners of all grades

- know algorithms to solve tasks and problems from different application domains and read and interpret given algorithms,
- design and realise algorithms using algorithm modules and present them in an appropriate manner.

*Languages and Automata (C3)*

Learners of all grades

- use formal languages to interact with informatics systems and to solve problems,
- analyse and model automata.

*Informatics Systems (C4)*

Learners of all grades

- understand the basics of the structure of informatics systems and how they work,
- use informatics systems in a targeted manner,
- exploit further informatics systems.

*Computer Science and Society (C5)*

Learners of all grades

- identify interactions between informatics systems and their social environment,
- exercise freedom of choice in dealing with informatics systems and act in accordance with social norms,
- respond appropriately to risks in the use of informatics systems.

The Process Domains describe the way in which learners deal with the subject content. They give indications of how the content can be didactically implemented in lessons.

*Modelling and Implementing (P1)*

Learners of all grades

- create computer models for given situations,
- implement models with suitable tools,
- reflect on models and their implementation.

*Reasoning and Evaluating (P2)*

Learners of all grades

- ask questions and make assumptions about computer science issues,
- justify decisions in the use of informatics systems,
- apply criteria for evaluating computer science issues.

*Structuring and Interrelating (P3)*

Learners of all grades

- structure facts by breaking them down and ordering them appropriately,
- recognise and use connections within and outside computer science contexts.

*Communicating and Cooperating (P4)*

Learners of all grades

- communicate computer science issues in a professional manner,
- cooperate in solving computer science problems,
- use suitable tools to communicate and cooperate.

*Representing and Interpreting (P5)*

Learners of all grades

- interpret different representations of facts,
- illustrate computer-related facts,
- select suitable forms of representation.

Since about 2009, a GI working group then dealt with the development of standards for computer science at the senior secondary level. These standards were adopted and published by the GI in January 2016, after several years of discussion in the expert committees (GI – Gesellschaft für Informatik e.V., 2016a). The structure of the junior secondary level was retained, but the description of the contents within the domains was slightly adapted (especially with regard to the aspect of Modelling). In addition, three requirement domains were distinguished as a third dimension for the senior secondary level in accordance with the requirements for the uniform Abitur examinations:

1. Reproduction
2. Reorganisation and transfer
3. Reflection and problem-solving.

This results in a three-dimensional model:

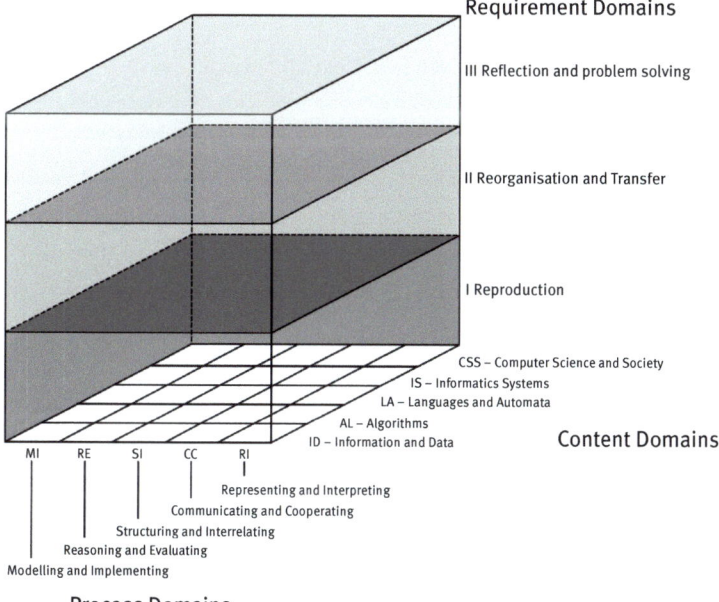

*Figure 24.* Competence model of the educational standards in computer science at the senior secondary level

The respective competence domains are differentiated in the standards, according to requirement areas and for levels EF and Q1/Q2, they are differentiated according to the requirement levels for a basic or advanced course. For a description of the content, please refer to the standards themselves.

## 2.5.2 Mapping the Components of International Curricula into the Framework of a Proposed Competence Model for Computer Science Education at the Primary Level

The competencies from the international standards and curricula analysed in the previous section are hereinafter presented within the frameworks of the GI standards for the secondary level. Here, it becomes apparent that the framework can largely be used to deal with the areas in the age range at the primary level, albeit to a different extent or with clear focal points. Furthermore, it becomes clear that the entire human-computer interaction interface area for ICT application competence and *digital literacy* within the computer science competencies can only be covered if an additional Process Domain is defined. This Process Domain, which we will hereinafter refer to as *P0: Interacting & Exploring*, is often found in the international standards.

In the appendix, the competencies mentioned in the international standards and curricula are summarised under the respective assignments to the Content Domains in the original. In the following, these assignments to computer science processes are summarised first and then the Content Domains.

**Classification of the International Standards in the Process Domains**

From the competence descriptions of the international approaches, it is often not directly clear how learners deal with the subject content. Therefore, in the following, an allocation of all described competencies to the Content Domains can be made and thus the possible Process Domains can be shown, since the way of dealing with a subject content often depends on the didactic discussion in the teaching situation. Often, these competencies are only referred to as "understanding" or "recognising". Children can actively gain understanding by modelling & implementing something (P1) and justifying & evaluating parts (P2) in the process. In doing so, aspects are often structured & networked (P3) and, depending on the applied didactic method, cooperated & communicated (P4). Subject content must also be presented and the subject presentation interpreted (P5). Above all, children need to actively engage with subject concepts, algorithms, programmes or informatics systems to develop structures and processes by interacting with and exploring them (P0) in order to build a mental model. Therefore, the authors of this expert report extend the GI competence model at the secondary levels to include this Process Domain.

*(P0) Interacting and Exploring*

After analysing and discussing the competence descriptions in the international standards, the team of authors proposes to introduce an additional Process Domain for the primary level competence model. This Process Domain describes basic competencies for exploring the use of informatics systems or components. How can one interact with informatics systems in a systematic and sound way? How can they be explored systematically in order to build up a mental model of modes of action, functionalities and structures? These basic skills must be acquired as a prerequisite to support the other processes of contextual learning. If they are acquired sufficiently at the primary level, they can be used and deepened at the junior secondary level. Therefore, they no longer appear in this model but are considered as an orthogonal competence in the other five Process Domains.

The Process Domain is often combined with Content Domain C4 (Informatics Systems) (cf. Section 3.4). In summary, such competencies are described in all international standards for the age levels considered: Use **informatics systems** or components and typical IT processes (C4) …

- in a sound manner (components, architecture, devices, document storage, etc.)
- purposefully/adequately (especially communication and cooperation in the creation of digital artefacts)
- to select and compare (prerequisite for justification and evaluation)
- to explore and build a mental model, inferring functionalities and explanatory internal structures from an external view of systems
- to learn with informatics systems

Although less detailed in the standards, all other Content Domains can always be combined with Process Domain P0 Interacting & Exploring:

- **C1 Information & Data:** Store and find data (files, names, types), use databases, search for information, develop a model of how to organise information as data, etc.
- **C2 Algorithms & Programmes:** e.g., apply simple algorithms (control a robot) while exploring step by step, act out algorithms while interacting with objects, explore programmes.
- **C3 Languages & Automata:** this Content Domain is hardly described in the international standards. However, examples can also be imagined in the young age group, e.g., entering a web address (URL) in the correct format, setting an alarm clock, ...
- **C5 Computer Science & Society:** recognise in which domains informatics systems play a role; this can be done e.g., by role-plays.

*(P1) Modelling and Implementing*
The **entire Content Domain C2 (Algorithms and Programmes)** is essentially (but not exclusively) linked to this Process Domain (detailed list in Section 'Classification of Content Domains' and in the appendix), e.g.,

- a pupil should know how to write executable **programs in at least one language.** (CAS, p.14, **Programming**)
- **Construct** a set of **statements** to be acted out to accomplish a simple task (e.g., turtle instructions). (CSTA p.14, L1:3.CPP 4.)
- **Implement** problem solutions using a block-based **visual** programming language. (CSTA p.14, L1:6.CPP 6.)

- can write and test **programmes** with loops, conditional instructions and parameters. (CH LP21 KL. 3-4)

*(P2) Reasoning and Evaluating*
Many of the Content Domains offer the possibility to reason and evaluate (see possible assignments in the following sections). However, the competence descriptions often only explicitly list the topics without showing in which way the understanding takes place. This can often include the possibility of reasoning and evaluating a fact, even though this will be rather the exception at the young age level, e.g.,

- C4 (Informatics Systems): Apply strategies for **identifying** simple hardware and software **problems** that may occur during use. (CSTA p.14, L1:6.CD 3.)

*(P3) Structuring and Interrelating*
Process Domain P3 is often an implicit part of P1, because modelling involves analysing and structuring an issue. Furthermore, modelling, representation and interpretation (P5) is about networking with computer science issues. Therefore, P3 can play a role in all Content Domains, especially in C2 (Algorithms and Programmes), certainly also in C1 (Data) and C4 (Informatics Systems). Especially in C5, networking with subject areas outside computer science plays a role. Examples found in the international standards are:

- (C1 Information and Data): Gather and **organize information** using concept-mapping tools. (CSTA p.14, L1:3.CPP 6.)

- C2 Algorithms (and Programmes): Algorithms are developed according to a **plan** and then **tested**. Algorithms are corrected if they fail these tests. (CAS, p. 13, Algorithms, Key Stage 2)

- C1, C2, C4 (Informatics Systems): Understand how to **arrange (sort) information** into useful order, such as sorting students by birth date, without using a computer. (CSTA p. 13, L1:3.CT 3.)

- C1, C2, C4, C5 – if applicable (Computer Science and Society): Make a **list of sub-problems** to consider while addressing a larger problem. (CSTA p. 13, L1:6.CT 5.)

- C5: Understand the **connections between computer science and other fields**. (CSTA p. 13, L1:6.CT 6.)

- C2, C4, possibly, C5: It can be easier to **plan, test and correct** parts of an algorithm **separately**. (CAS, p. 13, Algorithms, Key Stage 2)

*(P4) Communicating and Cooperating*
In general, Process Domain P4 is more a question of didactic method and a generic competence. Therefore, it hardly appears explicitly in the Content Domains of the international standards without specific examples with methodical implementation. However, it is mentioned that the computer science method of problem-solving (C1, C2, especially C4) is often cooperative. It is also explicitly mentioned that communication and cooperation tools are also used for cooperative problem-solving (C4, C5). Examples are:

- Gather information and **communicate electronically with others** with support from teachers, family members, or student partners. (CSTA p. 13, L1:3.CL 1.)
- **Work cooperatively and collaboratively** with peers, teachers, and others **using technology**. (CSTA p. 13, L1:3.CL 2.)
- **Use productivity technology** tools (e.g., word processing, spreadsheet, presentation software) for individual and collaborative writing, communication, and publishing activities. (CSTA p. 13, L1:6.CL 1.)
- Use online resources (e.g., email, online discussions, collaborative web environments) to participate in **collaborative problem-solving activities** for the purpose of developing solutions or products. (CSTA p. 13, L1:6.CL 2.)
- **Identify** ways that **teamwork** and collaboration can support problem-solving and innovation. (CSTA p. 14, L1:6.CL 3.)
- Use computing devices to **access** remote information, **communicate** with others in support of direct and **independent learning,** and pursue personal interests. (CSTA p.14, L1: 6. CPP 7.)

*(P5) Representing and Interpreting*
Representing and/or interpreting computer descriptions occurs in all Content Domains and often in combination with other Process Domains. P1 is hardly possible without P5.

- Information can be stored and communicated in a variety of forms, e.g., numbers, text, sound, image, video. (CAS, p. 16, Data, Key Stage 1)
- Use writing tools, digital cameras, and drawing tools to illustrate thoughts, ideas, and stories in a step-by-step manner. (CSTA p.13, L1: 3. CT 2.)
- Structured data can be stored in **tables** with rows and columns. Data in tables can be **sorted**. Tables can be **searched** to answer **questions**. Searches can use one or more columns of the table. (CAS, p. 16, Data, Key Stage 2)

**Classification of the International Standards in the GI Content Domains**

In contrast to the Process Domains, it is easier to classify the competencies addressed in the international standards in the content framework of the GI, since the contents are explicitly listed in each case.

*(C1) Information and Data*
The Content Domain of presenting information in the form of data is taken up in all proposals except the conservative New Zealand proposal. However, it is not a focus in the approaches. In the implementation of the CAS proposals in the English curriculum, the domain has additionally lost importance and is only mentioned marginally. In the CSTA, the respective competence is explicitly described.

In summary, some approaches describe how children in the lower levels understand that different forms of information (text, graphics, images, audio) can be represented and processed on computers in the form of binary data. Furthermore, they should be able to collect and digitally manage information with different devices (photos, scans, audio recordings, …).

In the second age group, errors in the presentation and ways of dealing with them are also discussed. An explicitly mentioned data structure is the table; the mentioned procedures are sorting objects (or representations) according to selected properties and searching and finding digitally managed information. An informatics system mentioned in the environment of the representation of distributed information is the World Wide Web.

*(C2) Algorithms (and Programming)*
The Content Domain of Algorithms (here supplemented by Programming because this domain is mentioned in one breath in many international curricula) forms a focus in almost all international developments. There are numerous targeted competencies and many worked-out examples.

At the youngest age level, an understanding should first be developed that algorithms and programmes are formulated in terms of unambiguous predefined instructions. Precise language is needed to describe them. These principles can also be observed in the children's living environment, independent of the implementation in informatics systems. The first programming concepts (sequence, possibly already repetition, structured breakdown) can be worked out with suitable tools or everyday examples. First, algorithms, such as systematic sorting by properties, can be carried out and explored. Children can learn to follow clear instructions according to rules (interpret simple algorithms).

In the second age group, more advanced programming concepts, such as variables, conditional branching and loops should be used. As a rule, programming algorithms should be introduced with programmable toys and visual programming environments. The need for programming languages (with pre-defined meanings and the possibility of unambiguous formulations) should be understood. Also, the process of systematic development (structured breakdown, use of digital tools for problem-solving, development and testing) should also already be practised.

*(C3) Languages and Automata*
In the actual sense of GI standards of the concepts of theoretical computer science (automata, grammars), there is naturally nothing on this Content Domain in the international approaches for lower age groups. There is a recent article about a lesson on automata theory for the "primary school", but it refers to experiments in grades 5 to 7 (Isayama, Ishiyama, Relator & Yamazaki, 2016). If the domain is interpreted as describing an understanding of the representation of languages and their properties, as well as rules and tools for status-driven programming, part of the aspects mentioned in the previous Content Domain C2 (Algorithms & Programming) could also be assigned here, especially when it comes to the control of robots etc. This domain is mainly considered in the UK approach, less so in the American or New Zealand approach.

In Chapters 3 and 4, we formulate some ideas for possible implementation (which are not found in this form in the international curricula), e.g.,

- formal structure of rules, e.g., of street names, groups in child-care centres, structure of URLs
- systematic exploitation and configuration of informatics systems as a "formal" language (P0).

*(C4) Informatics Systems*
The domain of Informatics Systems is again a focal point in all international approaches. It is often about recognising components and functionalities, being able to exploit them, applying them in a well-founded and reflective manner, while exercising the necessary caution (interface to the following Content Domain, 'Computer Science and Society'). A large part of the competencies associated with this Content Domain is closely linked to the newly proposed Process Domain P0

'Interacting and Exploring'. All in all, the relations to IT literacy and sound user skills are obvious here; skills that naturally form a focus in the young age groups and are a fundamental prerequisite for computer science skills to be developed further, unless they are pursued purely "unplugged".

*(C5) Computer Science and Society*
The Content Domain 'Computer Science and Society' is found in all international approaches with a focus on safe and ethically correct use of informatics systems, especially the World Wide Web and social media available on it. Children should first learn that it is important to consider to whom one discloses which data and behind which actions dangers may lurk. They should learn to turn to parents and teachers in case of doubt. On the other hand, they should understand that there are people at the other end of the communication channel and that communication rules therefore also apply when interacting with programmes.

The Content Domain also includes recognising how much everyday life is permeated by informatics systems. They should learn to identify informatics systems and perceive their roles.

## 2.6 Results/Conclusion

The studies on children in a digital world mentioned in section 2.1, show that while children are generally confident in their use of informatics systems, they are often unaware of the underlying principles and therefore cannot always properly assess the consequences of their actions. In addition, their media use is largely consumptive (using media offers on the Internet, watching videos, playing games). They neglect the possible constructive and creative opportunities to create something new with and within informatics systems and to shape their digital world in a meaningful way. In our opinion, a special role of schools and possibly also child-care centres could be to promote and point out further possibilities and types of interaction in addition to the predominant consumption activities. Moreover, interactions could not only be experienced in isolation and individually, they could be experienced by parents and children together, so that the experiences can be verbalised and processed, as well as reflected on and classified in an age-appropriate way.

The learning-psychological preconditions that have been established speak for the fact that computer science education in child-care centres and in primary school makes sense and is possible. Children are in principle cognitively capable of grasping, understanding and implementing selected concepts of computer science. They can be enthused about aspects of computer science, and this applies equally to boys and girls.

In addition, there are a number of tried and tested learning tools for different approaches that seem fundamentally suitable for use in child-care centres and primary schools. Accesses with and without computers, tablets and smartphones can be used. Often, programmable toys enable experiences in the creative, design and programming handling of technical devices, which can be explored and investigated in a playful way. Alternatively, micro-worlds and development environments with visual, block-oriented programming languages are available. These alternatives offer children the opportunity to actively explore phenomena of a world shaped by informatics systems at an early stage and to experience the first steps of a computer-science mindset.

So far, little research has been done on which competencies should be targeted in the young age groups as a pre-requisite for a continuous school-based computer science education. Nevertheless, some progressive international education systems are currently introducing computer science education as a continuous compulsory subject in primary and secondary education to provide learners with a computational mindset and problem-solving skills using computer science methods at an early age. This way of thinking qualifies the learners both for well-founded and reflective handling of informatics systems and for the use of informatics systems to creatively shape their own life-world. We have compiled the intended competencies of the principles, standards and curricula from Great Britain, the USA, New Zealand and Switzerland, analysed them in detail and placed them in the framework of the competence structure model for computer science education established in Germany (GI – Gesellschaft für Informatik e.V., 2008). An important finding of this analysis and classification was the extension of the given GI model by a further Process Domain, which describes the interaction and exploration of informatics systems and computer science methods.

The resulting extended competence model is used in chapters 3 and 4 to present the goal dimensions for children and educators in a structured way and to discuss them using specific examples and implementation proposals.

## 3 Goals at the Level of the Children

The following goals at the level of the children describe possible goal competencies that can be achieved by children in child-care centres and primary schools through appropriate measures. The structure of these goals is based on *Weinert's concept of competencies*. According to this, competencies are

> "the cognitive abilities and skills available or learnable by individuals to solve specific problems, as well as the associated motivational, volitional and social readiness and abilities to be able to use the problem solutions successfully and responsibly in variable situations" (Weinert, 2001, p. 27).

In terms of computer science competencies, these correspond to the necessary cognitive abilities and skills, which can be both specific to computer science and interdisciplinary and include motivational, volitional and social readiness and skills.

In order to represent competencies in a structured way, competence models, especially competence structure models, are used. Typically, this structure of a competence model is oriented towards the requirements to be fulfilled. The different "competencies and sub-competencies are primarily defined according to the content of the relevant tasks and the requirements to be met in order to solve these tasks" (Hartig & Klieme, 2006).

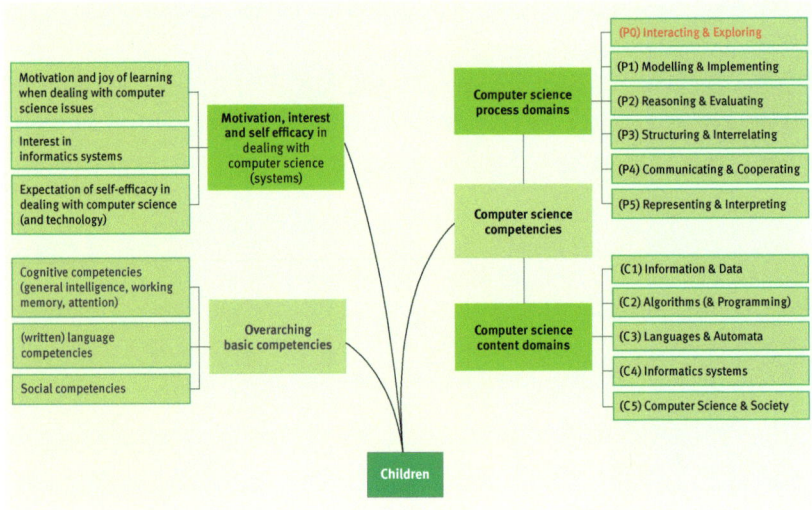

*Figure 25.* Goals of computer science education at the child-care centre and primary school level

The model presented below is based, among other things, on the model for early mathematical education formulated by Benz et al. (2017) and transfers overarching basic skills, motivation, interest and self-efficacy for dealing with informatics (systems), as well as computer science competencies (Benz et al., 2017). The latter is described in the form of a competence structure model, which, as explained in Chapter 2, is based on the GI recommendations for standards in computer science at the junior secondary level (GI – Gesellschaft für Informatik e.V., 2008). The draft for educational standards for the final stage of the senior secondary level, which was adopted by the Gesellschaft für Informatik concurrently with the preparation of this report, has also adopted this structure of Process and Content Domains. The proposal for GI recommendations for the primary level also adheres to the established structure of the first-mentioned standards (GI – Gesellschaft für Informatik e.V., 2008). From the authors' point of view, it therefore makes sense to try to adopt this structure largely for the elementary and primary level in order to ensure connectivity for the competence models of the secondary level. In addition, however, the international discussion and the international curricula on computer science education, especially at the primary and elementary level, will also be taken into account. The relevant results of our analysis in Section 2.4 of this expert report and fundamental considerations on a computer science exploration and design circle have prompted us to expand the established structure of the GI educational standards to include the process dimension P0 'Interacting and Exploring'.

## 3.1 Overarching Basic Competencies

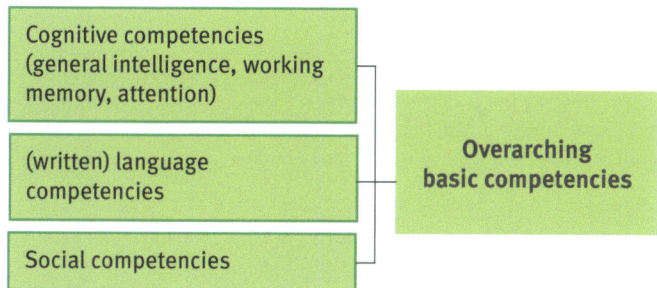

**Figure 26.** Overarching basic competencies

The basic competencies listed below are not the exclusive goal competencies of computer science education. On the one hand, they can influence processes of computer science education, while on the other hand, they can be promoted by learning activities within the framework of computer science education. This applies above all to the areas of problem-solving competence and social competence.

Since, with the appropriate level of competence, these subject-independent basic competencies can positively influence the computer science learning process, they may also be relevant as control variables in the accompanying research. The overarching basic competencies comprise three categories: cognitive, linguistic and social competencies.

### 3.1.1 Cognitive Competencies

Cognitive competencies in the sense of general intelligence or attention play a role in all types of learning. It can be assumed that correspondingly strong early influences also have an influence on the learning of computer science content, especially in the area of problem-solving. Cognitive competencies are therefore suitable as control variables within the framework of an evaluation.

### 3.1.2 (Written) Linguistic Competencies

The acquisition of computer science competencies is closely linked to the development of computer terminology, which includes numerous anglicisms (computer, hardware, software, …). Even children who do not know English can acquire these concepts if they are introduced to them by educators who serve as role models. Here, it is essential that the early childhood educators are acquainted with the terminology and use it consistently. Currently, there are no specific measurement tools to assess language development in computer science.

### 3.1.3 Social Competencies

Social competencies always play a major role when learning and working in a team or in groups. This is because this form of learning and working has a key function in computer science in particular, be it in the joint processing of problems, pair programming or the discussion of, for example, the evaluation of a system, software, etc. or the assessment of social and ethical framework conditions. Social competencies are always also a goal of computer science, since they are needed for the typical project-like forms of learning and working and are thereby promoted.

## 3.2 Motivation, Interest and Self-Efficacy of Computer Science

As presented by Benz et al. (2017) for early mathematical education, the psychological constructs "motivation" and "interest" are considered as closely related and grouped under the umbrella term 'motivation' (Anders, Berwanger, & Stiftung Haus der Kleinen Forscher, 2013a, 2013b; Benz et al., 2017; Schiefele, 2009):

**Interest** refers to a clearly delimited Content Domain (e.g., on football, butterflies, classical music or surprise eggs), whereas motivation aims at an action to be performed. Individual interest is a permanent, dispositional characteristic of the individual. For school age children, interests for individual school subjects (and thus Content Domains such as mathematics, social studies and science in primary education/natural sciences or technology/computer science or similar) can certainly be identified. For pre-school age children, the authors consider such specific interests to be rather unlikely, unless there are exceptional abilities and associated interests (e.g., in sports or playing a musical instrument).

**Motivation**, however, which in literature is often equated with "situational interest" (Seel, 2003) (e.g., to perform a certain action), is a common driving force even at pre-school age. But motivation is not triggered by interest immanent to personality, but arises in pre-school age from the play experience, from the stimulus of the game itself, as well as from the emotional closeness to the playmates. Especially for computer science lessons, this offers the possibility to deal with informatics systems and concepts in a playful way. A positive or negative attitude towards individual games or game types and forms is only likely to develop in the course of play experience, which is conditioned by the positive or negative experiences in comparative situations. According to the authors, situational interest in computer science is not to be expected at pre-school age but can develop at primary-school age if subjects are considered important (Benz et al., 2017).

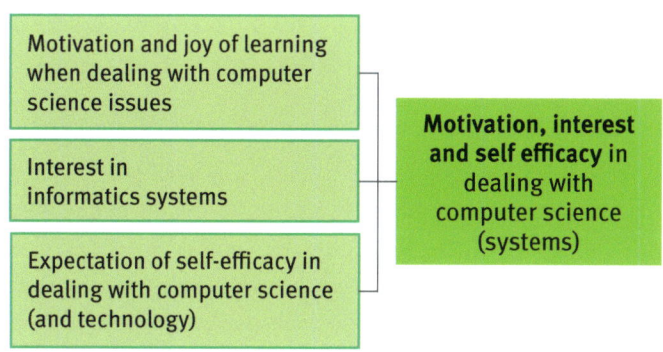

*Figure 27. Motivation and interest*

The constructs discussed in detail in Benz et. al. (2017), 'motivation', 'interest' and 'self-efficacy' can be assumed analogously for the subject of computer science (Benz et al., 2017). However, there are far fewer reliable findings from corresponding studies for computer science, since computer science basics have hardly been institutionally laid down at the pre-school or primary level so far.

There are some studies on self-efficacy in the context of pure computer use that compare the attitudes of pre-school children with those from primary school (grade 2): "Overall, children express positive attitudes towards computers (M = 3.23, SD = 0.56). Significant differences in attitudes (...) towards computers were found for gender (in favour of boys), grade level (in favour of the older children in grade 2) and socio-economic status (in favour of the children from a lower socio-economic neighbourhood, Southside)" (McKenney & Voogt, 2010, p. 16). Vekiri & Chronaki (2008) found similar results for primary schools.

The advantage of computer science is that it is always possible to engage with informatics systems age-appropriately in a playful way, since informatics systems and the concepts of computer science realised in them can be represented in the form of toys (cf. Section 2.3). Like the authors of the mathematics expert report, we believe it makes sense to engage in computer science in a playful way, e.g., through programmable toys (Benz et al., 2017). This mainly applies to pre-school age but also to primary school: "Basically, it can be stated for pre-school age that the playful approach is genuine for children, while training programmes, on the one hand, cannot develop motivational power and, on the other hand, are not more effective than the playful approach" (Benz et al., 2017, p. 47; Hauser & Rechsteiner, 2011; see also Pauen & Pahnke, 2008).

The authors are aware of only one study that examined the influence of extracurricular measures for computer science education on the development of self-efficacy. In his dissertation, Leonhardt found measurable differences between boys and girls with regard to their basic attitudes towards technology and computer science (Leonhardt, 2015). To compensate for these differences, he designed targeted didactic measures for girls aged 11-12 years and has empirically proven medium to strong effects of strengthening their self-efficacy. These measures consist of an intervention workshop (2-day robot programming courses for girls in their schools) and private follow-ups (3-day advanced programming workshops) during the holidays to stabilise the incipient individual interest. The empirical pre-study and post-study of the intervention workshop, together with that of the follow-up workshop, provided findings regarding the continuation and stabilisation of effects on the technical, computer-science self-concept as well as on expectation of self-efficacy and the conviction of control in dealing with technology. A stabilising *emerging individual interest* is evident among the participants after the follow-up workshop. There is a strong desire to deal with the subject matter again and again, i.e., the prerequisite for a stable personal interest in the subject matter exists. The intervention measure is a first step towards levelling the emerging gender difference in the self-concept in dealing with technology and computer science and shows that the gap between environmental socialisation taking place and measures promoting interest starting too late can be closed. The studies give reason to believe that early and continuous computer science education can prevent gender differences with regard to the expectations in self-efficacy and interest in computer science from developing in the first place.

*Recommended target aspects for computer-science self-efficacy*
For the development of a positive computer-science self-concept, it is beneficial that children have positive experiences when dealing with computer-science content. Tasks that are close to reality are suitable for this purpose; they must not overburden, but also not under-challenge. Through the playful handling of these tasks, children can expand their computer knowledge and gain experience with computer representations and regular processes. They discover that the concepts they have learned and their understanding of typical processes are suitable for organising and grasping the (virtual) world of informatics systems. This gives them confidence to explore and help shape the (digital) world. They feel the effectiveness of their actions, which increases their sense of self-efficacy. According to attribution theory, for feedback, successes should be attributed to one's talent in order to build or maintain a high self-concept, while failures should rather be attributed to a lack of effort or external factors (Leonhardt, 2015).

## 3.3 Computer Science Competencies of Children

In the following, the Content Domains and Process Domains for characterising computer science competencies at the level of the children are described. In this way, important goals for computer science education for this age group can be presented. This description is closely related to the corresponding goals for early childhood educators and primary school teachers. In section 4.3, the Content Domains and Process Domains are therefore revisited and specified, taking into account specific requirements for this target group.

*Content Domains* characterise the computer-science content to be acquired by the children. *Process Domains* describe the way in which children are expected to master this subject content (see GI – Gesellschaft für Informatik e.V., 2008, 2016a). In other words: Content Domains describe what children should learn about computer science, and computer science processes describe how computer science is to be "done/executed". The Content and Process Domains presented here result, on the one hand, from the educational standards for computer science and, on the other hand, from the analysis on which the previous chapter was based. There, the authors of this expert report came to the conclusion that it makes sense to supplement the computer science processes described in the GI recommendations for educational standards for computer science with the area of Applying and Exploring (P0). In principle, all Content Domains can be combined with all Process Domains.

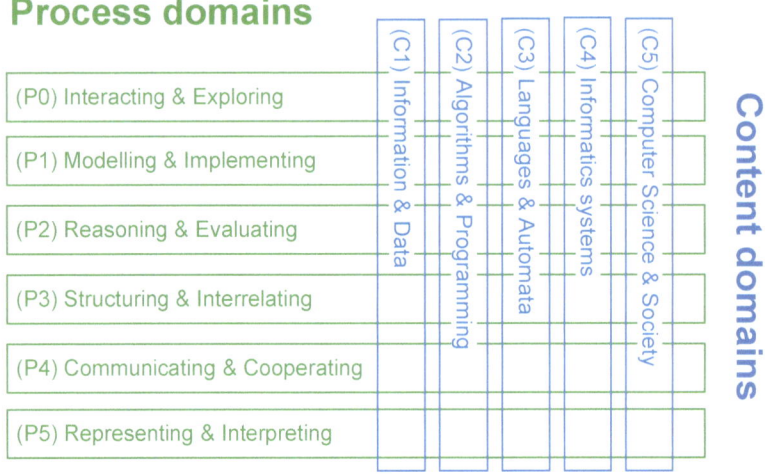

*Figure 28.* Interlinking Process and Content Domains for the representation of computer science competencies

### 3.3.1 Content Domains

**Information and Data (C1)**

In computer science, the systematic presentation and automatic processing of data as a carrier of information plays a central role. Information is understood as the context-related meaning of a statement, description etc., while data is the presentation of information in a formalised way. It is thus suitable for communication, interpretation and processing.

Data can code information in various forms of presentation (images, texts, symbols). A representation in a programme is in each case a string of characters that follows a fixed syntax in order to be automatically processed and interpreted. Examples that are likely to be familiar to children are Morse code, sign languages such as flag signals or Braille. At the level of computer science, binary code plays a major role. Data are represented as binary code, which in turn can be converted into on and off signals at the physical level.

Data becomes information when it is interpreted in a context of meaning. A familiar example from the learners' environment is letters and words composed of these letters (data) and the meaning of the words (information). Another example are pictures composed of e.g., many pixels.

Children should be able to understand the connection between information and data and distinguish between representational form and meaning. When translating information into certain forms of representation (data), e.g., SOS into Morse code, children carry out the process of representing or, conversely, interpreting (P5). Children can represent things according to properties they have chosen themselves (P1), e.g., to order them (P3).

**Algorithms and Programmes (C2)**

The term algorithm is a central concept in computer science. It describes specific instructions for action in a fixed sequence, which are necessary to solve a problem or achieve a goal. For the goals at the level of the children, this means that they encounter algorithms by solving and describing age-appropriate problems with the help of instructions for action. These can be simple tasks, such as finding a way through a maze, or more complex problems, such as sorting data.

In the children's world, rules of action can be found, for example, as rules for games, step sequences for tying shoes or brushing teeth, painting instructions, directions, cooking recipes, etc. They can read and interpret these rules in different forms of representation, e.g., graphically, as text or in a programming language (P5) and thus execute and explore them symbolically (P0). Furthermore, such instructions can be extended both colloquially and by own self-invented rules (P1).

Depending on the age of the children, they can already work out forms for representing algorithms. Instructions for action can be in the form of symbols (building blocks, cards, e.g., to control a robot) or later in the form of precise written instructions (link to C3). The introduction can be "unplugged", by using simple symbols on paper and putting them in the right order, e.g., to control a paper robot. In further stages, age-appropriate (e.g., visual) programming environments can also be used. Basic programming concepts, such as e.g., sequences, conditional instructions and loops/repetitions can be worked out with children (cf. children's website of "Haus der kleinen Forscher" Foundation http://www.meine-forscherwelt.de/, ScratchJr, AgentSheets). They can also find mistakes in given sequences or add to/complete them in order to solve a task. An important insight should be that clear and unambiguous formulations, as well as exact and similar execution is important. Furthermore, children should learn that they themselves can design processes in a virtual world by adapting and further developing algorithms (P1).

**Languages and Automata (C3)**

Languages are known as a means of communication between humans but also between humans and informatics systems (input and output), as well as between different informatics systems. In computer science, language is a formalised representation of information. Such formal language that follows a fixed syntax is the prerequisite for machine processing by automata (see GI – Gesellschaft für Informatik e.V., 2008).

Children know automata, for example, in the form of toy cars, ticket or beverage machines, smartphones etc. They are state-based systems that read and process an input. If a vending machine receives an input, it changes its state. In this new state, it can process information and then output something or wait for a new input, for example. Both would lead to a change of the state again.

Children learn that these automata can be controlled specifically by inputs and that automata are always in a certain state. To communicate about states of automata, graphical representations of state transitions can be used.

The goal of computer science education should be for children to understand automata from this computer science perspective. This means that they can recognise states and state transitions in everyday situations (P1) and, if necessary, also represent them graphically (P5).

**Informatics Systems (C4)**

It is impossible to imagine everyday life without informatics systems. Every washing machine or even coffee machine often already contains an informatics system. Even small children play with smartphones and tablets and make intuitive, playful experiences with informatics systems (P0), although they do not always recognise them as such (games console, smartphone, television). Informatics systems are specific assemblies of hardware and software components for solving one or more problems (P1) and, if applicable, input and output devices. This can already be the small toy robot that can be controlled with the help of different buttons or a doll that can be put into different modes via settings and thus play different music (reference to P1 and C2, partly C3). In contrast to the abstract concept of automata described in the previous section, informatics systems are concrete realisations that children can interact with to have concrete experiences (P0).

Informatics systems are often networked and communicate with each other via technical communication protocols (reference C3) or with other technical components of the system via actuators and sensors (embedded systems). 'Communication' with the users of the system is also important. This is done via user interfaces (e.g., display) that enable humans to 'communicate' with and control the informatics system. This is called 'man-machine communication'. This is an essential prerequisite for the efficient use of the system. Therefore, informatics systems are often also referred to as socio-technical systems, since the social action systems of interacting persons associated with the system must be taken into account when modelling and designing the system (Magenheim, 2008; Ropohl,

1999). Didactic concepts of computer science, such as the 'system-oriented approach' or 'computer science in context', attempt to take this social component of informatics systems into account in learning processes related to computer science (Koubek et al., 2009).

The construction and handling of these systems (reference to C5 and P0) is part of computer science education. What are the components of informatics systems? How do they interact? What functions are realised by which components? If problems occur, how can they be limited and solved?

However, computer science takes us even further behind the "scenes". How does the input and output work, i.e., how does communication come about in computer science (reference to C1 and P5)? Here, in particular, the input-processing-output principle, or IPO principle for short, plays an essential role. The input, which is made via an input device, is processed by the computer system. The system responds, for example, with an output on the monitor. It quickly becomes clear that, when looking at an informatics system, in addition to the hardware, the software also plays a role. There are thus different operating systems, often with similar basic structures and other software that already plays a role in children's everyday lives. Basically, children have to learn that informatics systems work with electricity (C1), how programmes are started and closed down without data loss (P0) and that logging into a system can play a role.

Another important aspect is networking of informatics systems up to the Internet. Its structure and functioning should also form part of the contents of computer science education. An understanding of the structure and functioning of the Internet supports, for example, the explanation of phenomena such as the freezing of videos on the Internet. It helps to understand whether data is located locally on one or several servers on the Internet. The question of whether data is located on one's own computer or on a server on the Internet is an essential basic computer science competence. The interconnectedness of informatics systems becomes clear when children, for example, send each other messages or use digital cameras, transfer pictures to computers and then process them further with graphics and drawing programmes. Child-friendly cameras have an editing function already integrated, which makes it clear that cameras are also computer systems.

An essential prerequisite for the development of IT systems is the competence to explore them systematically and to develop a suitable *mental model* of the systems (P0). In this context, the realisation that informatics systems are adaptive (P2) and thus suitable for creative action and expression plays an essential role. This enables the transfer of knowledge to different systems and lays the foundation for the competent use of informatics systems for the desired purpose. Related to this are questions of how children use systems safely and in which cases they turn to which persons to report events in the digital world. Finally, motor skills for

using gestures or input devices must also be learned.

**Computer Science and Society (C5)**

The increasingly far-reaching role of informatics systems in our world has a significant impact on our society. Some of these aspects should also be part of early computer science education. In this way, it is possible to discuss positive and negative social behaviour with children, also when dealing with technology and social media. Further ethical aspects, such as legality in dealing with informatics systems, can form part of computer science education. How can I protect my personal data? Why do I need to protect it in the first place? Why should I use an envelope when sending mail? These are also important aspects of dealing with informatics systems that should be taught as early as possible.

Furthermore, the role of informatics systems in everyday life can be discussed. What is the significance of these systems for our daily routine (especially that of the children)? What impact can these have? What significance do they play in the professional world and in which professions does computer science play a role? Children should recognise how the world is permeated by informatics systems. They should understand the principles of computer science on different informatics systems, such as camera, smartphone or toys and thus understand the basics of the progressive integration of media and systems (TV, smartphone).

An important competence is first to learn to distinguish between people and informatics systems and to identify when people are involved through the medium of informatics systems and when a programme is involved (P2). Last but not least, connections should be drawn from computer science to other sciences such as electrical engineering or physics (P3).

### 3.3.2 Process Domains

For our competence model, we adopt the five Process Domains mentioned in the GI proposals for educational standards for the junior secondary level and senior secondary level, as well as in the GI draft for the primary level, in order to enable a direct connection to the established competence models for the secondary level:

**(P1)** Modelling and Implementing
**(P2)** Reasoning and Evaluating
**(P3)** Structuring and Interrelating
**(P4)** Communicating and Cooperating
**(P5)** Representing and Interpreting

As justified in Section 2.4, we add the additional Process Domain

**(P0)** Interacting and Exploring

This additional Process Domain illustrates that the basis of competent use is systematic and active exploration.

### Interacting and Exploring (P0)

Playfully exploring and trying out objects in their everyday world is an essential element of children's leaning processes (see Sachser, 2004). They literally 'grasp by grab' such objects and acquire a rudimentary mental model of their technical functions and possible 'inner life' through playful interaction with them (see Schwarzkopf & Zolg, 1997; Zolg, 2006). The findings of learning psychology on how children deal with simple technical artefacts can also be used for learning with and about age-appropriate computer systems (e.g., Bee-Bot, …). By interacting with and exploring informatics systems, children can gain initial access to these systems. If these experiences can be systematised for children in moderated learning processes, they can create the basis for a deeper insight of the children into the function and structure of such systems.

In addition to the learning-psychological justification of this Process Domain, there are also numerous subject-didactic and subject-scientific arguments that justify the relevance of this Process Domain for computer science education. Numerous international approaches for early computer science education, e.g., Lifelong Kindergarten at MIT (Resnick, 2013) or the CAS initiative in Great Britain (cf. Section 2.4), contain 'Tinkering', the playful exploration of informatics systems, which is a methodological concept comparable to the Process Domain P0 'Interacting with and Exploring of Informatics Systems'.

In the subject didactic discussion, there are numerous approaches that attach great importance to the exploration and use of informatics systems in computer science education. In the application-oriented (Körber & Peters, 1988) and in the use-oriented approach (e.g., Forneck, 1992) of computer science didactics, the reflected use and application of informatics systems is seen as relevant for learning processes in computer science education. In the system-oriented didactic approach, the externally visible function and the internal structure of informatics systems should be systematically explored by means of the method of deconstruction (Magenheim, 2008). Similarly, in the subject didactic approach to duality reconstruction, the learners' everyday experiences with informatics systems are to be linked to, and the duality of structure and function of digital artefacts is to be explored through systematic use (Schulte, 2009). In Stechert's subject-didactic approach, too, the exploration of informatics systems should allow conclusions to be drawn about their inner structure from their externally visible behaviour (Stechert, 2009).

Finally, subject-didactic competence research in computer science has identified the ability to explore and use informatics systems as an essential prerequisite for understanding them (e.g., Magenheim et al., 2010).

From a subject-specific perspective, the constant technical change in the field of informatics systems necessitates that both experts and laypersons interact with and explore them. Laypersons should be enabled to understand the functioning of informatics systems by means of user-friendly interfaces and, if necessary, to configure them according to their specific needs. This form of competent use of informatics systems is thus increasingly becoming part of general education.

When developing new informatics systems, experts usually first explore an already existing system and its context of use in order to recognise its basic functioning and design principles. Based on these findings, new system components can then be constructed or the system as a whole can be further developed.

From a general educational perspective, exploring and using informatics systems opens up the possibility for children to build up a mental model of informatics systems that can be systematically expanded and redefined in the course of further learning processes. This also puts them in a position in the future to deal with new software or new technologies and, if necessary, with new interaction patterns, and to understand technical changes in a reflected way as an opportunity and a risk. In this way, children are enabled to act competently in the course of their lives, also in the use and application of future IT systems, and to derive a benefit from new technologies.

Of course, the competencies of this Process Domain can hardly be realised "unplugged" because it is precisely about the interaction with informatics systems (C4), whose use is to be observed and about whose behaviour assumptions can be made (P1, P5), which then have to be validated.

**Modelling and Implementing (P1)**

Modelling and Implementing are the central components of the software development cycle and thus a core process of computer science. In software development, the initial situation and a problem therein is analysed, usually several computer science models are designed with an increased degree of formalisation, which are finally implemented, tested and reflected on an informatics system. All other Content and Process Domains are usually closely linked to this.

In **modelling**, relevant aspects of reality or those of a planned informatics system that is to change reality are represented by abstraction for a specific purpose. For this purpose, a requirements analysis is carried out to examine facts and processes from a computer science perspective. The aim is to describe the essential and generalisable components and parameters of a system, as well as their interrelationships as clearly as possible. In additional modelling steps, further increasingly formal, textual and/or graphical models are designed by a structured break-down of the initial problem, the identification of technically representable subcomponents and the construction and networking of the components into a system architecture.

Through **implementation**, a formal computer science model is described on an informatics system (usually in one or more programming languages) in such a way that it can be executed on such a system. In addition to the actual programming, this process step also includes the systematic testing and assessment of the created solution according to functional and non-functional criteria (evaluation), as well as the reflection of the system in the context of use (validation), which evaluates its possibilities and limitations and often leads to new requirements and a next iteration of the modelling and implementation cycle.

At the level of children in elementary and primary education, the aspects of professional software development cannot be fully taken into account. However, it is possible to break down the initial problem into smaller problems (C4, P3) and to identify components, depending on the task. Initial problems or tasks come from children's experience. The same applies to the structured presentation of solutions (C2, C3 and also P5). The aim is not to formally describe informatics systems for solving tasks, but to adopt computer science perspectives and use modelling steps for everyday tasks and for describing everyday phenomena. In addition, children should acquire the competence to "read" (e.g., C3) and interpret (P5) simple computer science (often graphical) models in order to solve a task. Furthermore, they should reflect on their own or given model descriptions. How does the behaviour change when a part of the model description is changed? How must a description change in order to achieve a certain effect?

The first steps of implementation can also be carried out in special, child-friendly working environments (ScratchJr, MicroBits etc., cf. Section 2.3). The focus is especially on programme creation and its methods and ways of thinking. These can already be included in computer science education at the elementary and primary level. Small programmes can be created or finished programmes can be adapted by implementing algorithms in a suitable, child-friendly (usually visual) programming language. In addition, these can be easily checked for correctness, as the products created can be experienced and compared with one's own expectations of the system. A great advantage is that informatics systems themselves provide feedback to the children. Implementation can also take the form of physical elements, such as blocks or cards (unplugged). However, executing one's own implementation is then no longer possible. The system itself can then no longer provide feedback and can no longer be checked and validated for correctness as easily.

### Reasoning and Evaluating (P2)

As a rule, there are fundamentally different ways of solving tasks or problems using computer science methods. Therefore, computer science contexts, procedures and approaches to finding a solution must always be explained and justified. Only in this way, for example, can different approaches, solutions or systems be compared with each other and evaluated according to different criteria. This requires that technical language is learned together with general communication and argumentation skills and should lead to the reflective use and adaptation of informatics systems.

Children should justify decisions appropriately, assess computer science issues according to values of the subject and form their own opinions. This requires knowledge of the terminology, rules, methods and procedures, as well as an understanding of computer science. Reasoning involves the conscious consideration and weighing up of the advantages and disadvantages of different computer science approaches (C2) or representations (C1). Evaluating presupposes one's own position with regard to computer science facts on the basis of appropriate evaluation criteria and standards (C5).

For the children, this means that they learn by means of specific examples to ask questions and make educated assumptions about computer science contexts and to check these on the informatics system (also part of P0). With the help of simple criteria, they can evaluate computer science facts.

### Structuring and Interrelating (P3)

When structuring, individual components of facts or processes must be recognised and related to each other. This refers to various areas of computer science. For example, information can be organised and data can be structured for specific access and processing (C1), sub-problems can be constructed and/or broken down (C2), or sub-functions or components can be described (C4). Various kinds of connections can be made within and outside of computer science (C5). The creation of planning processes also falls into this area. In computer science, complex requirements are often processed using computer science tools, which include structuring and interrelating in particular.

For children, this means that they playfully learn the first steps of breaking down facts or processes, with the aim of structuring and solving tasks in partial steps. First arrangements of everyday objects can be made on the basis of criteria, e.g., assigning children to learning groups or classes. Vice versa, children can explore structures and connections using examples of their everyday world.

**Communicating and Cooperating (P4)**

This Process Domain is usually a matter of didactic approach and is thus interdisciplinary. Nevertheless, it plays a special role in computer science education. On the one hand, communication can take place via computer science contents. On the other hand, communication can also take place with the help of an informatics system as a medium. Joint work can also be conducted without an informatics system, e.g., when working together on a computer science problem in a team. But also here, for example, the joint work on a text can take place with the help of an informatics system.

Selected according to age, children can get to know/deepen/expand the various possibilities of communicating and cooperating. Multi-touch devices and learning programmes adapted to these devices can play a special role. As a competence in computer science education, children should learn to represent their thought processes when solving computer science tasks in colloquial language and increasingly also with terms of a technical language in order to understand them vice versa. This forms the basis for collaboratively solving computer science tasks.

**Representing and Interpreting (P5)**

In computer science, there are various graphical and textual forms of representation of an informal, semi-formal or formal nature for different problems and approaches. At the level of the children, age-appropriate forms of representation should be chosen, which the children can use and also interpret themselves. In doing so, they can formulate their ways of thinking and proceeding when solving computer science tasks. This can be done verbally, by drawing or in writing, using different types of representation such as graphs (objects with connections), tables or lists etc. They learn to explain computer science facts that are presented graphically or in simple, age-appropriate formalisms (telling stories, reproducing processes). For this, age-appropriate learning environments can be used (Storytelling Alice, ScratchJr.).

## 3.4 Prioritisation of Specific Competence Expectations at the Level of the Children

In the following, goal competencies at the level of the children are listed, which, in our view, show the most important age-appropriate links between Process and Content Domains. This prioritisation is based on guiding criteria that are oriented both to the subject-didactic relevance of the competence area to be selected and to learning and developmental psychological criteria. In addition to the subject-related computer science competencies, which result from a combination of Content and Process Domains, and into which, above all, subject-didactic considerations flow, the aspect of teaching general basic competencies and the contribution to children's general education that can be expected with this, plays an important role in the recommendation of prioritisation.

All in all, it should be noted that these recommendations were developed primarily from the perspective of computer science didactics on the basis of the currently still quite limited research situation. The perspective of elementary and primary pedagogy and didactics on this subject area is only just emerging, so that the suggestions described in the following should be understood as impulses for testing and for developing an educational field of 'computer science for the elementary and primary level'.

The following considerations also play a – not insignificant – role in the compilation of criteria for setting priorities: Although computers have played a role in media education in child-care centres and primary schools for quite some time already, this is primarily a matter of appropriate, critical and reflective media use (Gesellschaft für Didaktik des Sachunterrichts, 2013, p. 9; Senatsverwaltung für Bildung, Jugend und Wissenschaft Berlin, 2014, p. 103). However, the central task of social studies and science in primary education is to "support learners to understand their natural, cultural, social and technical environment in a factual way, to explore it on this basis in an educationally effective way and to orientate themselves, participate and act in it" (Gesellschaft für Didaktik des Sachunterrichts, 2013, p. 9). Without computer science ed-

ucation, however, these goals cannot (or can no longer) be achieved in a world shaped by computers.

Social studies and science in primary education therefore urgently need innovations on the topic of 'computers': In order to support children in understanding their technology-influenced environment and to prepare them adequately for the future in this respect, the broaching of the subject of so-called 'new media' must go beyond application learning and also beyond the teaching of general media skills:

> "Within the computer science community (academia [sic!] and industry) there is general agreement that computer science is not about how to use a computer through the applications that it can execute, but it is about knowing how these applications work" (Gibson, 2012, p. 34).

This goal is in line with the requirements for competence development in social studies and science in primary education (see Gesellschaft für Didaktik des Sachunterrichts, 2013, p. 63), so that the future inclusion of computer science education in the educational areas of the primary level, especially in social studies and science in primary education, seems only logical and appropriate with regard to technical and social developments.

The reasons why computer science education beyond media use has so far only had a low status in primary schools are manifold and similar to those that, in the past, have led to a rather marginal engagement with natural science and technical content (see Döbeli Honegger, 2010; Köster, 2006, p. 11). Reasons for the rather marginal engagement may be personal fear of contact or an attitude of avoidance on the part of early childhood educators and primary school teachers but also a lack of competence with regard to computer science content, as well as topics and challenges perceived as more urgent in everyday school life. Therefore, when implementing IT education in primary schools, inhibition thresholds or personal 'teaching limits' (Köster, 2006) must be taken into account, which can lead to children experiencing 'learning limits', even though they themselves are interested in content of computer science education.

In summary, the selected competence areas are prioritised on the basis of the following guiding criteria:

1. **Subject-specific and subject-didactic concepts:** What is the significance of the chosen competence area for the understanding of central concepts of computer science and their application in informatics systems?
2. **Learning and developmental psychological aspects:** Can concepts of computer science be taught at a cognitive level appropriate for the target group?

3. **Relevance to daily life:** Can the competence area be presented in an age-appropriate way in relation to the children's everyday situation?
4. **Motivation:** Can the prioritised competence area be presented using an example that is motivating for children?
5. **Professional interest:** Are the prioritised competence areas suitable to arouse the children's interest in computer science?
6. **Self-efficacy:** Does the competence area and the chosen example enable the children to strengthen their self-efficacy in dealing with informatics systems?
7. **General education:** Can the competence area and the chosen example contribute to children's general education?
8. **Overarching basic competencies:** Which basic competencies, such as communication and cooperation skills, empathy, problem-solving skills, are promoted by the chosen competence area?
9. **Reference to didactic concepts** (primary school, social studies and science in primary education, early childhood education): Can the prioritised competence area be placed in the curricular context of primary school and in concepts of early childhood education?
10. **Existing practical experience:** Does the selected competence area already show positive experiences in the pedagogical implementation of target group specific learning scenarios?

With regard to the term of general education, we are guided by the concepts of Klafki and Heymann:

Klafki's concept of general education includes (Klafki, 1993):

A1: Ability to co-determine
A2: Self-determination ability
A3: Capacity of solidarity

General education means

A4: Education for all
A5: Comprehensive education
A6: Education in the sense of 'general'

Heymann's concept of general education is defined by (Heymann, 1997):

A7: Preparation for future life situations
A8: Foundation of cultural coherence

A9: Building of a world view
A10: Guidance for the critical use of reason
A11: Developing a responsible approach to the competencies to be acquired
A12: Strengthening the learners' ego

### 3.4.1 Explanation of the Prioritised Competence Areas

The Content and Process Domains can and should be combined. The following table shows the individual possible combinations of a Content Domain with a Process Domain.

| Content Domains / Process Domains | (C1) Information & Data | (C2) Algorithms & Programming | (C3) Languages & Automata | (C4) Informatics systems | (C5) Computer Science & Society |
|---|---|---|---|---|---|
| (P0) Interacting & Exploring |  | LG |  | DG |  |
| (P1) Modelling & Implementing | LG | DG |  | LG |  |
| (P2) Reasoning & Evaluating |  |  |  |  | DG |
| (P3) Structuring & Interrelating |  |  |  | LG |  |
| (P4) Communicating & Cooperating | LG |  |  |  |  |
| (P5) Representing & Interpreting | DG | LG |  |  |  |

*Figure 29.* Combinations of Content and Process Domains. Dark green for combinations that are important and obvious to the target group. Light green: Combinations, for which we suggest further examples. (These examples are presented in chapters 4 and 5)

All in all, to us, the following combinations of Content and Process Domains seem to be particularly promising for successful computer science education at the elementary and primary level (highlighted in dark green in Figure 29):

- Modelling and Implementing (P1) of Algorithms and Programmes (C2)
- Interacting with and Exploring (P0) Informatics Systems (C4)
- Representing/Presenting and Interpreting (P5) Data and Information (C1)
- Reflecting on and Evaluating (P2) the Interrelationship of Computer Science and Society (C5)

For a practical example, there are usually also references to other Content and Process Domains. This means: In principle, in specific examples, several Content Domains and several Process Domains are involved in varying degrees. The primary purpose of the following prioritisation therefore is to focus.

The prioritisation is first done for the target group of children. It will later be carried out analogously for early childhood educators and primary school teachers.

In the following, we will first work out and justify what distinguishes the four selected combinations and why we consider them to be particularly important.

**Modelling and Implementing (P1) of Algorithms and Programmes (C2)**

This area can be summarised – slightly simplified – as programming. However, programming is a controversial term, as there are different understandings of what counts as programming.

In the 1940s and 50s, programming problems and tasks almost always referred to mathematical problems. In the 1970s, programming was seen as a "fundamentally easy task": It was thought that one only had to learn the (constructs of the) programming language and one could start (engage in). Programming courses were therefore almost pure language courses in which syntax elements were presented one after the other. Then, since the end of the 1970s and the beginning of the 1980s, programming has increasingly been understood as a complex problem-solving process, which, on the one hand, is about the (technical) solution to the problem – and, on the other hand, it is also about understanding the problem in the first place (Van Merrienboer & Krammer, 1987).

Consequently, two activities (and skills) have been distinguished since then: 1) designing (or modelling) and 2) implementing. In the first step, the problem is analysed and an algorithmic solution is designed; this is the design phase. In the second step, the solution is transferred into a programming language; the implementation or coding phase (see Van Merrienboer & Krammer, 1987). Since then, the term programming has often been used in different ways (cf. Figure 30).

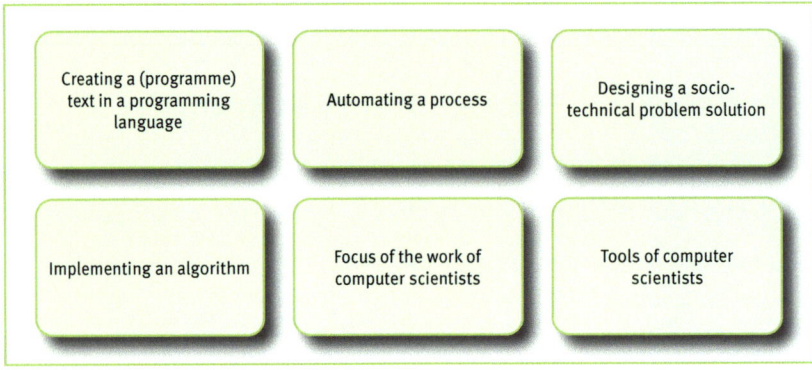

*Figure 30.* Views on programming

In Germany, a rather narrow understanding seems to be prevalent: The German-language Wikipedia focuses the term programming on the implementation phase: "Programming (from Greek prógramma 'prescription') refers to the activity of creating computer programmes. This includes above all the conversion (implementation) of the software design into a source code and – depending on the programming language – the translation of the source code into the machine language, usually with the help of a compiler".[39] The English-language Wikipedia defines the term even further: "Computer programming (often shortened to programming, scripting, or coding) is the process of designing, writing, testing, debugging, and maintaining the source code of computer programmes".[40] These two definitions reflect the distinction between design and implementation mentioned above. We understand the term in the broader sense (for more details, see Schulte, 2013) and include modelling, implementation, algorithms and programming language.

The aim is to understand, retrace and be able to develop simple sequences of action. This ability is not only useful for the children in the context of computer science education but can also be transferred to other educational areas in which problem-solving skills are required.

Examples of competencies expected in this area are:

- Dealing with rules of action (especially designing, but also reading and comprehension)
    - Children name and formulate action instructions for controlling an age-appropriate informatics system (also P5).
    - Children explain and read instructions and procedures for controlling an age-appropriate informatics system (also P4).
    - Children design a rule for encrypting messages (data) with age-appropriate procedures (e.g., Skytale).

- Implementing (use of a formal notation)
    - Children design instructions/sequences of action using given age-appropriate building blocks or commands.

- Simple algorithm (applying and investigating rules of action)
    - Children apply the given procedure to find a faulty part (also P5).
    - Children explain given algorithms.
    - Children execute algorithms step by step (simulate, also P5).

---

39  http://de.wikipedia.org/w/index.php?title=Programmierung&oldid=113866966
40  http://en.wikipedia.org/w/index.php?title=Computer_programming&oldid=537629774

**Interacting with and Exploring (P0) Informatics Systems (C4)**

Playful exploration and trying things out as an essential element of children's learning can be implemented in particular by combining Process Domain P0 in relation to informatics systems and their modes of operation. In this process, children should gain experience in order to establish the basis for a deeper knowledge of the function and structure of such systems.

Handling a digital artefact is only the basic level of this competence area. The focus is on system exploration and goal-oriented interaction, with which the informatics system or an algorithm is opened up. Exploration also includes adaptation and creation, in the sense of end-user design. Knowledge of concepts of computer science makes it possible to open up different views of an informatics system or a digital artefact. Since not all relevant properties and states of an informatics system can always be read directly from the user interface, this is an important prerequisite for the system understanding to be developed.

Interaction as a competence is not limited to handling the system, but includes thinking about the meaning of interaction. This can go beyond the individual user and the individual use case and relate to a group of people, at a basic level, e.g., when choosing a tool for use by one's own group in the child-care centre.

Children should not only learn how to specifically deal with a digital artefact but also acquire and trust general and transferable strategies to explore an unknown system, while also thinking about its possibilities, limits and effects.

An important aspect of this "exploratory competence" is the insight that most artefacts allow for adaptation and adjustment in relation to one's own wishes, i.e., they can be designed by oneself. This then leads to the possibility of being able to make these adjustments and to have the confidence to do so. This can involve configuration, parametrisation up to smaller programming activities in the sense of end-user programming.

Above all, children should realise that computer science is not only interaction, i.e., dealing with a system, but also designing and realising ways of dealing with it.

This includes, for example, the following competencies:

- Exploration (investigating, trying things out with the aim of gaining knowledge about the digital artefact)
    - Children discover simple functions of particular programmes or websites and find them again with repeated use. (also P3)
    - They identify informatics systems in their environment and how they can use them to their advantage. (also P2)

- The children learn to draw conclusions about the internal structures and processes and states of an informatics system, by observing it from the outside. (also P2 & P3)
- Children apply strategies to identify simple hardware and software problems that can occur during use.

- Interaction (recognising, using and adapting paths of interaction)
  - Children use input and output devices in an age-appropriate and confident manner, especially audio instructions and gestures on touchscreens and special symbols.
  - Children use technical tools (such as multimedia applications, text editors, web tools, etc.) to present age-appropriate tasks. (also P4)
  - Children navigate between websites by using hyperlinks and construct simple queries using search engines. (also C3)

- Raising awareness (evaluating the artefact or a path of interaction)
  - They learn typical patterns of informatics systems, e.g., the arrangement of icons and selection menus and recognise personalised use after logging on to a system. (also P3)

### Representing and Interpreting (P5) of Information and Data (C1)

Central to this area is the concept of information, which gives the discipline its name – and yet this is difficult to grasp. According to Breier, information is the "meaning of a statement, instruction, notification, message, communication or the like" (Breier, 2004a, p. 69). In relation to computer science education, Breier lists some essential characteristics of information (Breier, 2004a, p. 69):

- Information does not exist 'in itself', but must be represented on a carrier medium.
- For this purpose, a code is agreed between the sender and receiver of the information (syntax and semantics).
- If the representation is destroyed, the information may also be deleted.
- Information can be processed by processing its representation.

These characteristics show that a strict distinction is made between two aspects of a message: the meaning content, on the one hand, and the form and the representation, on the other (see also Hubwieser, 2007, p. 78). The latter is the representation of a message – (only these) representations can be processed by the computer. The former, i.e., only the meaning content, is the information. It is (only) accessible to humans or more generally: biological organisms (cf. the following figure).

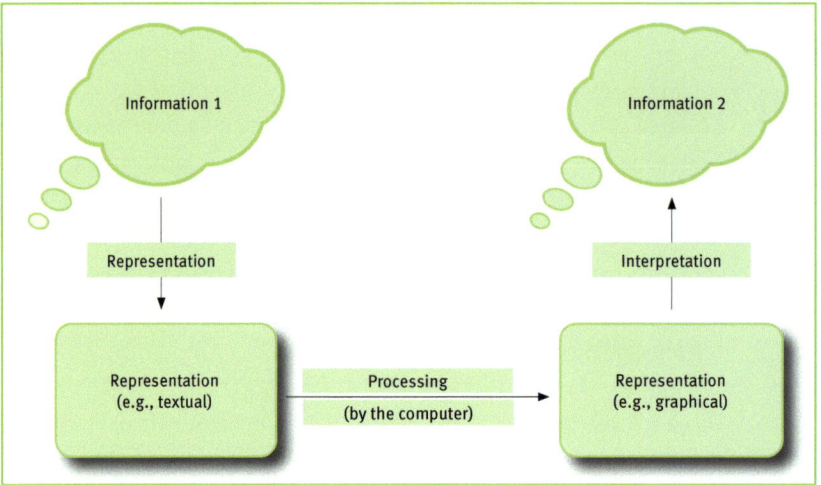

*Figure 31. Schematic representation of information processing (according to Breier, 2004a, p. 74, Hubwieser, 2007, p. 80, or GI – Gesellschaft für Informatik e.V., 2008, p. 23)*

The representation of information can be understood as data – strictly speaking, computers therefore only ever process data, not information. According to the scheme, humans gain information by interpreting the changed representation or the changed data.

The competence area Representing and Interpreting Information and Data therefore deals with different forms of representation of data or different representations of information. An important question in this context is, for example, how information is transformed into digital data and how this data can be transmitted and interpreted. A specific question is, for example, how pictures get onto the monitor or "into the computer".

Examples of specific expected competencies are:

- Represent information (digitise): Children ...
    - are able to explain how information is converted into digital data.
    - can convert decimal numbers, letters and graphics into binary numbers and vice versa.
    - can represent (P1) and order (P3) things according to self-selected characteristics, so that they can find an object with a certain characteristic (e.g., colour, form, size) more quickly.

- collect data on a task and represent it by using child-appropriate computer science tools (editors, digital cameras, drawing tools, concept maps) to construct thoughts and stories step by step.
- learn to represent answers to questions as a series of yes/no decisions.

■ Process data: Children …
- recognise that data can be represented, stored, processed and transmitted by an informatics system in different forms, e.g., as texts, pictures, videos, audio, combinations of these, …
- analyse the steps involved in encrypting messages (data) with age-appropriate procedures (e.g., Skytale).

■ Interpret data: Children …
- can recognise and identify errors in binary data.
- can explain that the Internet contains a very large amount of data.

**Reflecting on and Evaluating (P2) the Interrelationship of Computer Science and Society (C5)**

One reason for introducing computer science lessons in early childhood is the changes that digitalisation is bringing about in many areas of life: in the so-called effects of technological progress or the penetration of the life-world with digital artefacts – and in the scientific discipline that provides the knowledge and skills to develop these artefacts. But what is the relationship of computer science, human beings and society?

Often, only a general connection is assumed, when discussing this field: Computer science and society or technological progress as the driving force of societal change – but such technological determinism cannot adequately explain the changes. Rather, there seem to be complex interactions between computer science, people and society.

An interesting attempt to structure these interactions has emerged in the project 'Contextual Computer Science'. The core idea is the following: "Technology development is expressed not only in the respective artefacts, but also in what we call socio-facts (written and unwritten laws and agreements) and cogni-facts (following Foucault's 'technologies of the self', i.e. competences, methods and thus also techniques in the original literal sense)" (Engbring & Selke, 2013, p. 113).

According to this approach, not only new digital artefacts (=technical products) but also other "products" are created in the process of techno-genesis: At the level of society, new rules, laws, ways of interacting with technology (and, through this, also the interaction between people) emerge – and at the level of the individual, new sources of knowledge, skills and ideas emerge. These different processes and products penetrate and influence each other – and influence the process of genesis of new artefacts and technologies and thus the shaping of everyday life or the life-world.

This view of techno-genesis initially simplifies and structures the process but is neutral in terms of content. The authors propose to look at the areas work, culture and knowledge from this perspective and, finally, in the thematic field of Computer Science and Society, also to examine the techno-genesis in the three areas of a) **work processes** (development of machines and tools), b) **communication media** (changes in the area of culture/cultural techniques: writing, computing, media, communication and cooperation) and c) **knowledge society** (instruments and services for dealing with knowledge).

This results in a selection of contents and their structuring, see the following figure:

| Techniques | Design/Development | Regulation/Design | Exploration/Regulation | Emergence of technology |
|---|---|---|---|---|
| **Work** Machines Tools | Participatory system development | Occupational health and safety/ergonomics | Vocational training | Working processes |
| **Culture** Writing Arithmetic | Design of interactive media | Data protection/ informational self-determination | (general) education/ schools and universities | Communication media |
| Media Communication Collaboration | | Netiquette Telecommunication laws | | |
| **Knowledge** Instruments Services | System development as an adaptation | Patent Law Copyright | Professional societies/ responsibility | Knowledge society |
| | "Computer science in context" | | "Context of computer science" | |

*Figure 32. Proposal for structuring the basic area (according to Engbring & Selke, 2013)*

Furthermore, we propose to also use historical references for this purpose. The techno-genesis can be illustrated by individual examples: For example, how word processing with increasing layout possibilities and the advancement of printers changed the job profile of the typesetter/printer and how users increasingly design and produce the layout of their printed products themselves.

This area thus primarily deals with social, legal and ethical aspects in connection with computer science. Examples of competencies expected in this area are:

- Work/work processes: Children …
  - can name advantages and disadvantages of the use of robots or other informatics systems, depending on their age.
  - know steps in the development and digitalisation of writing systems.

- Culture/communication media: Children …
  - can find differences in interaction/communication with an informatics system or with a human being.
  - can explain how they should handle their personal data on the Internet.
  - can name the effects of cyber-bullying, depending on their age.
  - can set up rules for use for dealing with social networks and evaluate rules, depending on their age.

- Knowledge/knowledge society: Children …
  - recognise that informatics systems model "intelligent" behaviour.

- General/introductory: Children …
  - name advantages and disadvantages of the penetration of everyday life by informatics systems and tools, e.g., voice messages, playing or downloading videos, Internet access on mobile devices, navigation systems etc.
  - name the areas in their environment in which informatics systems are used.

## 4 Goals for Early Childhood Educators and Primary School Teachers

Teachers' professional competence and their job-related attitudes are essential factors influencing the quality of teaching and learning opportunities. This has been shown not least by Hattie's broad empirical meta-study (Hattie, 2009). Competence models for teachers have been developed for various subjects, including STEM subjects, as a theoretical basis for the empirical analysis of teaching quality, based on a concept of professional competence developed by Shulman (1986, 1987). The essential components are subject-specific knowledge (content knowledge – CK), pedagogical-psychological knowledge (pedagogical knowledge – PK) and, as a combination of both, subject-didactic knowledge (pedagogical content knowledge – PCK). Carlsen has modified Shulman's approach by dividing teachers' curricular knowledge into a subject-specific and a cross-curricular component (Carlsen, 2002). A good overview of the further development and differentiation of PCK models is given by Magnusson et al. (1999) and Fernandez (2014). In the subjects of physics (e.g., Riese, 2009), biology (e.g., Rozenszajn & Yarden, 2014) and especially mathematics (e.g., Lindmeier, 2011), these generic concepts have each been specified from the perspective of subject didactics in a series of studies such as MT21 (Blömeke, 2008; Blömeke, Kaiser & Lehmann, 2011), TEDS-M (Döhrmann, Kaiser & Blömeke, 2010) and COACTIV (Kunter & Baumert, 2011), and some of them have been empirically tested.

Models for describing the professional competence of early childhood educators and primary school teachers for the subject of computer science and computer science education hardly exist and have so far only been empirically verified to a limited extent (Hubwieser, Magenheim, Mühling & Ruf, 2013; see e.g., Saeli, 2012). In this respect, the research situation on the professionalisation of early childhood educators and primary school teachers in the subject of computer science is more comparable to the situation in technical education (Kosack et al., 2015) than to that of the subject of mathematics (see Benz et al., 2017).

The model of professional competence by Kunter & Baumert (2011), to which other concepts also refer (e.g., Döhrmann, Kaiser & Blömeke, 2012), serves as a starting point for the description of goal dimensions for early childhood educators and primary school teachers.

According to this model, important aspects of professional competence are 'motivational orientations', 'convictions, values and goals', the ability to 'self-regulate' and, on a cognitive level, 'professional knowledge'. The latter is subdivided into subject knowledge (CK), subject didactic knowledge (PCK) and pedagogical-psychological knowledge (PK). In addition, organisational and counselling knowledge are important for activities in the pedagogical context.

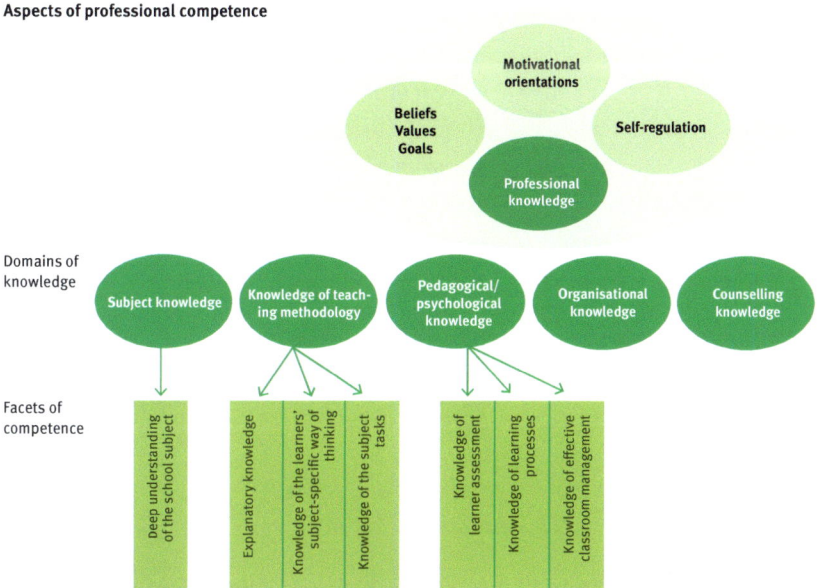

*Figure 33.* Aspects of professional competence (according to Kunter and Baumert 2011, p. 32)

Motivational orientation describes the disposition and perseverance of teachers to devote themselves to their professional activity. Self-regulation describes the ability to adapt one's own pedagogical behaviour in a situation-based manner to the respective specific framework conditions of the learning context, as it can be characterised, for example, with categories of the 'Hamburg Model' (Schulz, 1997). Finally, for the professional competence of educators and teachers, their understanding of their own role image as a teacher, the associated convictions and value attitudes towards teaching, as well as the significance of the subject they are to teach for the education of children and young people are highly relevant for their pedagogical practice.

This generic model of teachers' professional competence was developed primarily in relation to classroom teaching. In this respect, it must be adapted with regard to the organisational and content-related framework conditions to educators and teachers at primary schools and in the elementary sector and specified for the subject of computer science and computer science education. This will be done in the following sections of this chapter, as far as the research situation allows.

Section 4.1, first looks at the motivation, interest and self-efficacy of early childhood educators and primary school teachers. Section 4.2 is dedicated to the attitudes, approach and understanding of the role of pedagogues. The importance of the pedagogical-psychological knowledge (PC) of the educators and teachers with regard to the design of computer science-related learning processes in childcare centres and primary schools is discussed in section 4.2 as well. In sections 4.3 and 4.4, the required subject-specific competencies (CK) and subject-didactic competencies (PCK) of educators and teachers for designing learning situations for computer science education in the elementary and primary sector are presented. Finally, interdisciplinary aspects of learning with digital media are discussed from a computer science perspective.

The following diagram provides an overview of the goals of computer science education for early childhood educators and primary school teachers, the individual components of which will now be described and justified in more detail.

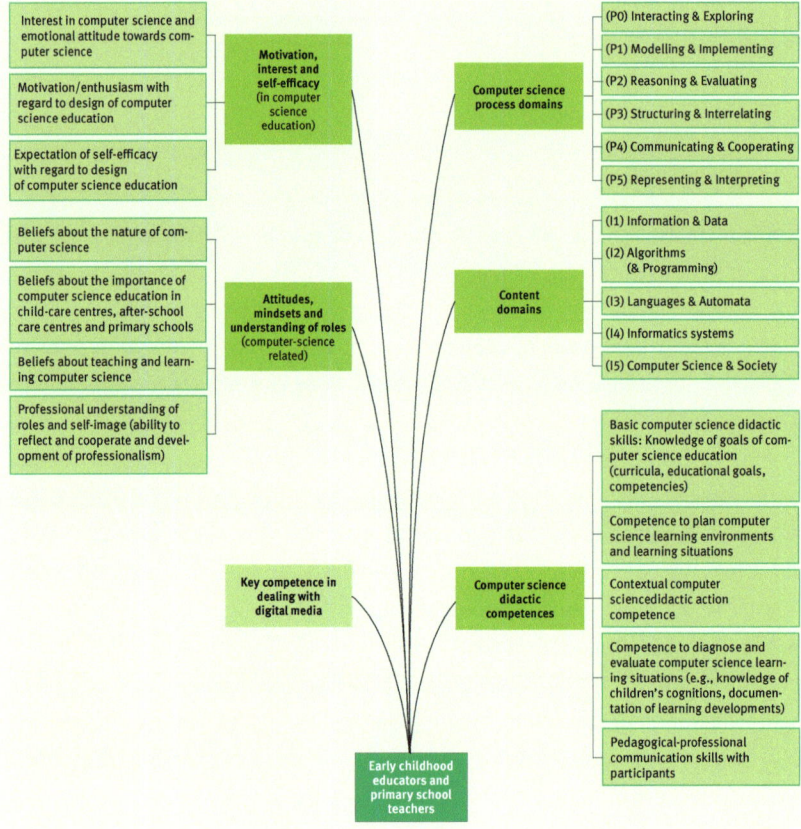

**Figure 34.** *Goals of early childhood computer science education for educators and primary school teachers*

## 4.1 Motivation, Interest and Self-Efficacy

According to Baumert and Kunter, teachers' motivation and interest in their subject and their concept of self-efficacy are of considerable importance for their pedagogical activities (Kunter & Baumert, 2011). For computer science education, there are so far mainly empirical findings on the level of learners, but only a few for teachers.

In the DFG research project MoKoM (Modelling Competence Measurement) (Magenheim et al., 2010), a theoretically based, subject-specific competence model was therefore developed, which, following Weinert's concept of competence, also includes non-cognitive competence dimensions such as motivational and volitional elements (Weinert, 2001). In a broad empirical study with computer science learners at the senior secondary level, the relevance of these facets for the competence level of the test persons could be proven (Neugebauer, Magenheim, Ohrndorf, Schaper & Schubert, 2015). In various other smaller empirical studies, the influence of technology acceptance according to the Technology Acceptance Model (TAM) (Bagozzi, 2007) on the use of an informatics system in a learning context was demonstrated, for example, among students of computer science/business informatics (Beutner, Kundisch, Magenheim & Zoyke, 2014). At the level of female learners, an empirical study was able to demonstrate positive effects of self-efficacy in a learning scenario in which female learners had positive experiences with informatics systems and their design (Leonhardt, 2015). The results of the study were presented in detail in Chapter 3. In a study with grade 2 learners, Kind was able to show that dealing with informatics systems and their programming (ScratchJr) had a positive influence on the learners' self-efficacy and could contribute to overcoming gender-specific differences in dealing with informatics systems (Kind, 2015).

The importance of self-efficacy concepts for teaching practice and the professional identity of "computer science teachers" as well as their consideration in concepts of computer science teacher training were highlighted (Ni & Guzdial, 2015). Within the framework of the German Federal Ministry of Education and Research (BMBF) research project KUI, Bender et al. come to similar conclusions in their evaluation of empirical data on the competence of female computer science teachers (Bender et al., 2015). The data from the study also provide indications of the importance of intrinsic motivation with regard to computer science topics and their implementation in the classroom. The authors specify Baumert and Kunter's professional competence model for computer science teachers by identifying the teachers' personal attitudes and beliefs and their motivational orientation – also with regard to perceived self-efficacy in computer science teaching – as an important component of competence (Kunter & Baumert, 2011).

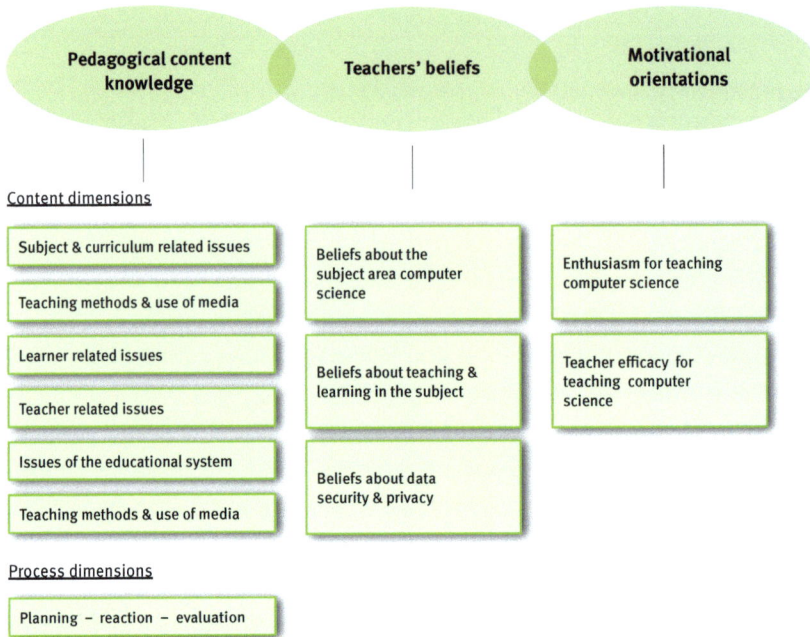

*Figure 35.* PCK for computer science teaching (Bender et al., 2015, p. 3)

For the goals of early childhood educators and primary school teachers in computer science education, it can be concluded from these results that both their own active experience with informatics systems and their medial use in learning situations with children, organised by the educators and teachers, should be part of the training.

## 4.2 Attitudes, Approaches and Understanding of Roles

The importance of teachers' attitudes, approaches and understanding of their professional role for their pedagogical practice was demonstrated in various empirical studies a long time ago. A good summary can be found, for example, in Eulenberger (2015). At the level of personality traits, the "Big Five" were used: These are the dimensions neuroticism (N), extraversion (E), openness to experience (O), agreeableness (V) and conscientiousness (G) (see Eulenberger, 2015, p. 1). Anders et al. (2017b) describe important components of the role and self-concept of early childhood educators and primary school teachers with the categories 'ability to reflect', 'openness', 'investigative attitude', 'ability to cooperate' and 'development of professionalism, among other things, through willingness for further training'. These general facets of competence certainly also apply to early childhood educators and primary school teachers in computer science education.

Kleickmann has investigated the connections between teachers' perceptions and teaching practice for natural science teaching in primary school, which could also be of importance for computer science education in terms of teaching methods (Kleickmann, 2008):

> "The greater degree to which primary school teachers hold an understanding that learners in primary natural sciences lessons learn best from the teacher's explanations and that learners tend to absorb knowledge passively-receptively ('transmission'), and that practical activities ('hands-on-activities') such as conducting experiments are a sufficient condition for learners to achieve conceptual understanding ('practicalism'), and also that learners in primary natural sciences lessons should work or learn largely independently, without the need for process-related support and structuring measures by the teacher ('laisser-faire'), the less progress the learners make in their understanding of scientific concepts" (Kleickmann, 2008, p. 172).

In a case-based long-term study of primary teachers' attitudes towards "effective science teaching", Davis shows that, although these attitudes are relatively stable, over time, they tend to move away from "reform-oriented" concepts to more "conservative", teacher-centred conceptions of teaching, based on the feedback from teachers' practical experience (Davis, 2008).

Fulton, in a study of junior secondary teachers' 'beliefs', had earlier pointed out that teachers' attitudes towards teaching styles dominate their use of digital teaching technology:

> "Results suggest that technology use did match teaching beliefs. Teachers with constructivist teaching beliefs adopted technology for a learner-centred teaching style, while those with more traditional (non-constructivist) teaching beliefs used technology in a more teacher-centred transmission style. Teachers said that technology has not changed their pedagogical beliefs, but the opportunities they have had to work in this learning community also affected teaching practice" (Fulton, 1999, p. 1).

Gil, Schwarz and Asterhan have investigated the influence of teachers' 'beliefs' regarding the use of digital media in learning situations and have shown that teachers' cognitive communication styles (intuitive moderation, synchronous discussion organising, guiding, observing, involved and authoritative) in these situations depend on their 'beliefs', their 'technical skills' and the 'constraints' of the learning environment (Gil, Schwarz & Asterhan, 2007).

Schulte and Bennedsen demonstrated the influence of 'beliefs' on computer science teachers' teaching practice in the context of object-oriented programming courses, and Fessakis and Karakiza in a study among Greek computer science teachers (Fessakis & Karakiza, 2011; Schulte & Bennedsen, 2006).

Ni specified the influence of 'beliefs' on computer science teachers' teaching practice in her study on the professional identity of 'CS teachers' (Ni, 2011). In addition to the general influencing factors, i.e., "perception of CS" and "perception of teaching", she determined the criteria of "self-identification", "interest/appreciation of teaching CS", "confidence in teaching CS", "learning styles to teach well", "retention of teaching CS" and "affiliation/belonging". She attaches particular importance to being well embedded in an educational institution and in a professional social network, which can certainly be used in Wenger's (1998) sense as a 'community of practice' for the acquisition of competencies in the sense of connectivism (Siemens, 2008).

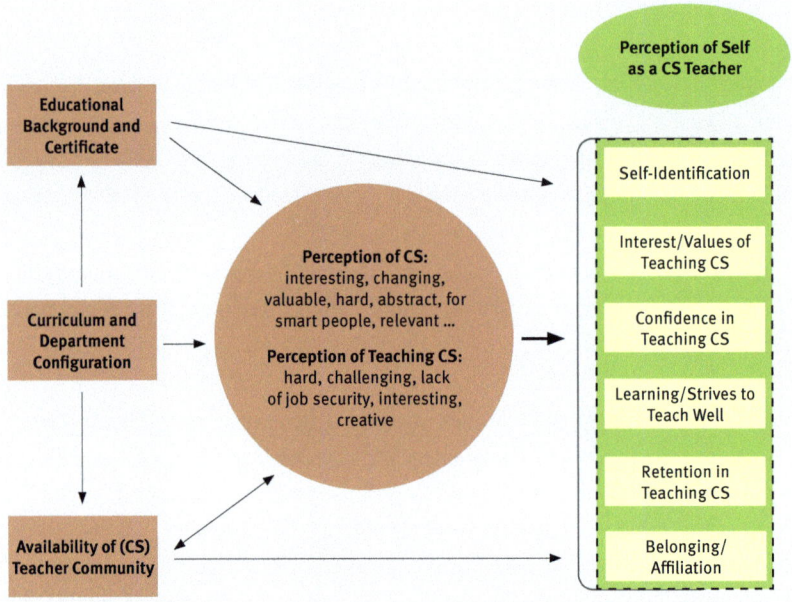

*Figure 36. Main Influencing Factors for CS Teacher Identity Formation (from Ni, 2011, p. 77)*

Finally, Bender et al. in the KUI project come to the following summary assessment regarding the influence of non-cognitive structural elements of competence (Bender, Schaper, Caspersen, Margaritis & Hubwieser, 2016):

> "The described theoretical analysis leads to assumptions that two main areas of beliefs (beliefs about the subject and its teaching and learning,

and beliefs about data security and privacy) and two main areas of motivational orientations (intrinsic motivational orientations and teaching efficacy) are particularly important for teaching computer science" (Bender et al., 2016, p. 6).

Reflecting on the possible influence of teachers' attitudes towards learning media used, especially digital media for supporting learning on their pedagogical behaviour in learning situations of computer science education, should be considered as an element of the objectives, as should embedding the discourse and exchange of information about this pedagogical activity in social networks of a 'community of practice'.

## 4.3   Computer Science Competencies

Various models and recommendations can be used to describe computer science competencies expected of early childhood educators and primary school teachers. However, these have been developed primarily for computer science educators and teachers working in out-of-school settings or at higher levels of school education. They can therefore only serve to a limited extent to concretise goals for early childhood educators and primary school teachers in elementary and primary education. Since most of the more recent recommendations use subject-specific competence models, these should at least be referenced here and related to the model we use. The GI recommendations for the design of Bachelor's and Master's degree programmes provide an up-to-date competence-oriented description of the subject-specific competencies of computer scientists at the Bachelor's and Master's level (GI – Gesellschaft für Informatik e.V., 2016b). The scope and level of the expected competencies described there admittedly go far beyond the required competence expectations for the target group of early childhood educators and primary school teachers addressed here. However, parts of these competencies will be incorporated into the competence expectations for computer science teachers at general education schools. The Standing Conference of the Ministers of Education and Cultural Affairs of the Länder (Kultusministerkonferenz der Länder) (Kultusministerkonferenz, 2015) has recommendations for these computer science teachers in the subject-specific and subject-didactic areas that cover the following Content Domains, which have been significantly reduced compared to the GI recommendations of 2016:

- Formal languages and automata
- Algorithms and data structures

- Data modelling and database systems
- Programming and software technology
- Computer structures and operating systems
- Computer science and society
- Subject didactics

There is a strong correlation here with the Content Domains of the GI educational standards (GI – Gesellschaft für Informatik e.V., 2008, 2016a).

While the GI recommendations for computer science degree programmes (GI – Gesellschaft für Informatik e.V., 2016b) contain a grading according to competence levels in accordance with an adapted AKT model (Krathwohl, 2002; Bröker, Kastens & Magenheim, 2014), there are no such gradings in the KMK recommendations and the GI educational standards. Other forms of grading subject-specific computer science competencies can be found in Fuller (Fuller et al., 2007), in the SOLO taxonomy (Biggs & Collis, 1982) or in the empirically based MoKoM competence level model (Neugebauer et al., 2015). Empirical studies show that elements of these competence level models and the subject content described above can be found both in the subject-specific training curricula of computer science bachelor's degree programmes at universities (Brabrand & Dahl, 2009; Müller, 2015) and in the subject-specific parts of the training curricula for students of computer science teaching (Hubwieser et al., 2013).

In the context of the objective of this report, however, it makes little sense to use these competence level models, some of which are quite complex, to describe the competencies of early childhood educators and primary school teachers in the elementary and primary sector, since neither a theoretically sound competence structure model nor empirically verifiable level models derived from it have yet been developed for this target group. This requires further, sometimes lengthy, empirical research.

With regard to the objectives for early childhood educators and primary school teachers in terms of the subject-related components of the competencies to be acquired, we are initially guided by the GI educational standards for the junior secondary and primary level (cf. also Chapter 3). In addition, further subject-specific arguments and subject-didactic criteria are taken into account. With this approach, we are also in line with the corresponding recommendations in the expert report for the subject of mathematics (Benz et al., 2017). The acquisition of subject-specific competencies in computer science should be closely related to the learning scenarios to be organised later for the children, so that the acquired knowledge can be quickly integrated into the children's own pedagogical activities. Therefore, the goals for early childhood educators and primary school teachers, their selection and prioritisation are also significantly oriented towards dimensions for children presented in Chapter 3. The interests and abilities of the children are also the focus of the methodological implementation in learning scenarios. Following the GI educational standards, computer science competencies are generated from the integration of contents and processes, as already presented at the children's level, by describing the respective expected competencies in a specific contextualisation of the learners' computer science process action in a Content Domain (cf. Chapter 3).

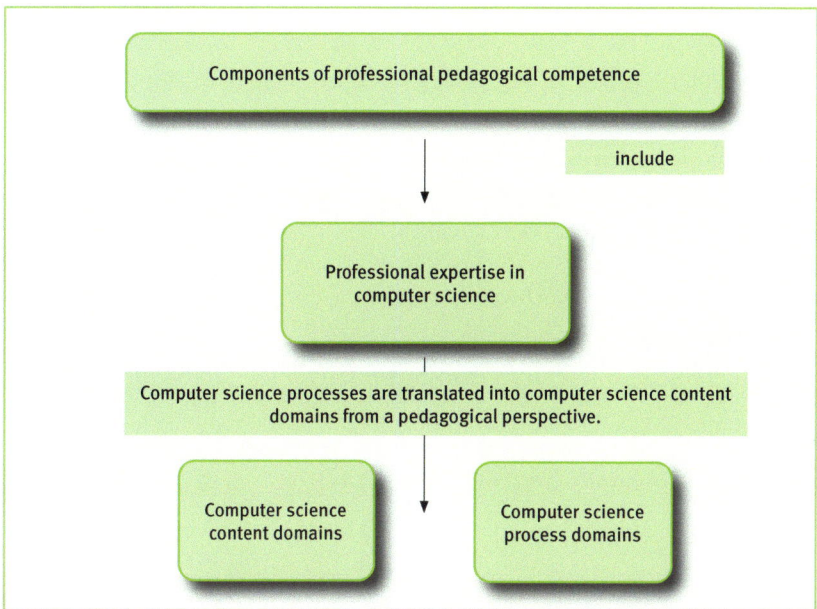

*Figure 37.* Professional pedagogical competence components in computer science education

The contextualised computer science process steps taken by early childhood educators and primary school teachers always occur in learning scenarios with children from the perspective of their professional pedagogical competence. Therefore, in the following descriptions of the Process and Content Domains, reference is always made to the corresponding competencies expected from the children (cf. Chapter 3), and the expectations regarding professional pedagogical competence in these learning contexts are discussed.

### 4.3.1 Content Domains

**Information and Data (C1)**

The ability to distinguish between information and data is an important content-related facet of competence that early childhood educators and teachers must have if they want to design appropriate computer science learning situations for the children. Different symbolic forms to represent data (e.g., letters, numbers, signs...) should be known, which can also be used in learning situations with children. It is also important to understand the transformation of information into data and vice versa, which takes place through the mechanism of assignment of meaning and interpretation.

For example, the sequence of letters, 'CAB', can be seen as an acronym in Latin or Cyrillic notation, or represent the number 3243 in hexadecimal. Crucial to the interpretation of the data is an appropriate interpretive context. Educators and teachers should be familiar with basic concepts of binary coding of data (numbers, letters, pictures, graphs) that enable their automatic processing and storage in a computer. An understanding should be developed that...

- different operations can be performed on the data depending on the data representation of a real-world object.
- the data representation of a real object involves processes of abstraction and decontextualisation.
- these processes are irreversible in some cases.
- different representations are differently suited for certain operations.

In addition to the examples for children documented in Section 3.4 or Chapter 5, educators and teachers can, e.g., use the following representations to illustrate this Content Domain: The aerial photo of a landscape – a cadastral map of the same landscape – the abstract representation of the map based on cadastral designations and plot dimensions – a data set with these plot data, whereby one can determine the size of the plot from the latter but not the vegetation of the plot. Pictures of a pizza, a cake or a house, for example, can serve as a starting point for further examples that are closer to the children's everyday world.

**Algorithms and Programming (C2)**

The concept of an algorithm, which is essential for computer science, is also an essential Content Domain for designing competence expectations for the early childhood educators and primary school teachers. It can be developed in a playful way, oriented towards the scenarios to be explored with the children, such as directions, rules of the game, pantomime, treasure hunt, cooking recipes or painting instructions. Here, the vagueness of everyday algorithms ('take ½ cup of milk and some butter') should be worked out and contrasted with the requirements for an algorithm that is to be executed precisely by an operator: Unambiguous instructions, finite sequence of instructions, termination of the programme. In this context, the essential control structures of an algorithm, such as sequence, conditional branching and loops/repetitions are learnt, just as an elementary variable concept with elementary data types. For the organisation of computer science processes, it is an important prerequisite that early childhood educators and primary school teachers master these basic concepts and can verbalise them accordingly to the children.

By playfully implementing an algorithm concept in learning scenarios, the operator can initially be a human being who follows the instructions of a "programme" and e.g., controls a paper robot (unplugged version). Later, the operator can also be a computer that executes a programme on the screen that was first developed in the form of puzzle pieces on paper and then entered as a computer-readable programme code, e.g., with the visual programming language Scratch. Here, a virtual robot is then moved on the screen in a similar way (virtual version). Finally, the same programme in the same programming language could also be used to control a small robot in the real world (real world version). It seems obvious to start the introduction to algorithms with an 'unplugged' version. For early childhood educators and primary school teachers it is advisable to get to know the other versions of algorithmic control because of the effects of self-efficacy and motivation described above (see also Section 4.1).

### Languages and Automata (C3)

Formal languages and automata are of great importance in computer science. For example, the theoretical concept of the Turing machine, an abstract automaton that is central to computer science, can be used to test the executability of an algorithm and describe its complexity. For early childhood educators and primary school teachers, it cannot be a matter of understanding these central concepts of theoretical computer science in detail. However, they should understand the basic principles of these concepts in order to be able to implement them in pedagogical learning contexts with children in an age-appropriate way.

For example, they should understand the difference between a natural language and the formal language that is independent of the 'interpretation context' and can be used, for example, to write a computer programme. References can be made to (C1). With regard to the concept-dependency of a natural language, communication models can be discussed with the early childhood educators and primary school teachers, such as, e.g., symbolic interactionism (Blumer, 1969).

Such models may already be familiar to early childhood educators and primary school teachers from their pedagogical training. In contrast, the concept of formal languages can then be elaborated in its diversity. This can be done in a playful and unplugged way. For example, the game 'Chinese Whispers' (or 'Telephone') can be used to illustrate the contextuality of natural language. Basic principles of a formal language can be taught in the context of communication protocols through communication by means of Morse code (drums) or light signals, whereby communication via light signals can also be done by means of binary code. If necessary, principles of encryption or rules for 'secret languages' can also be used here. The transmission of 'emails' can also be playfully explored in a role play by means of 'routing protocols' as a representation of a formal language.

In connection with (C2), programming languages, such as Scratch with its keywords and fixed syntactic rules for the design of instructions, can also be learned and applied as formal languages in an elementary form.

The starting point for dealing with the computer science concept of automata can be real automata that educators and primary school teachers encounter in everyday life: ticket machines, cash dispensers, traffic lights, vending machines etc.

The difference between a vending machine as an abstract theoretical concept and an informatics system as its possible technical realisation must be made clear to early childhood educators and primary school teachers. The invisible abstract theoretical concept of a vending machine can be understood, for example, through the visible input and output possibilities of an informatics system (e.g., cash dispenser). Here, too, early childhood educators and primary school teachers can get to know essential elements of the concept of an automaton by means of a role play: Inputs and outputs with corresponding systems, internal states, transition rules for states in the form of tables and graphs.

It is important in the playful approaches to the concepts of 'formal languages' and 'automata' that they are also formally described and understood in their essential properties at the end of the game, e.g., with a graph. In this way, a reference to an important concept of modelling (P1), state-oriented modelling, can be established at the same time.

### Informatics Systems (C4)

It is important for early childhood educators and primary school teachers that they can identify technical artefacts of the real world (phenomena of computer science) as informatics systems, to describe their essential visible and non-visible characteristic and also to understand them as specific technical realisations of abstract automata. On this basis, they are then also able to organise learning scenarios that introduce children to important properties of informatics systems.

Informatics systems have long been an integral part of our everyday lives, without us possibly even noticing them as such. We encounter them in the household (washing machine, dishwasher, microwave, coffee machine...), in communication and leisure (smartphone, tablet, laptop, television, digital media recorders, game consoles...), on the road (car, aeroplane, parking guidance system, GPS system, traffic lights...) or when shopping (scanner checkouts, ATMs, ticket machines, exit control systems, lifts...).

A detailed description of the concept of the socio-technical informatics system with its technical and social components has already been provided in Chapter 3. Informatics systems are usually very complex and difficult to understand, even for experts. However, a basic understanding of the structure and functioning of informatics systems is an essential aspect of computer science education.

Early childhood educators and primary school teachers should be able to experience these basic principles using simple examples. A first approach to informatics systems would be to identify the technical systems in the everyday world of children and early childhood educators and primary school teachers as socio-technical informatics systems, to describe some input possibilities, system reactions and possible actions of the users.

The examples should be closely oriented to those concepts that are also used in learning scenarios with the children. It should be possible to have both the external and the internal view of an informatics system. The external view can be achieved by analysing user interfaces (GUI: Graphical User Interfaces) and the associated man-machine communication (MMC), e.g., by analysing familiar software such as frequently used apps or websites with regard to input options and feedback. Optionally, it could be considered whether early childhood educators and primary school teachers first design the structure and layout of a website unplugged from an MMC point of view, taking into account the separation of structure, layout and content, and then implement it with simple HTML commands as an example of a mark-up language.

With the help of the IPO principle (input-processing-output principle), it is then possible to infer internal processes of the system and analyse the feedback behaviour (e.g., with apps like WhatsApp or Google). Internal states of the system, the processing of inputs and generation of outputs, which are described with algorithms, among other things, can be modelled with playful concepts using simple informatics systems (e.g., Lego robots, computer games) and then implemented with a visual programming language such as Scratch (cf. Chapter 3).

Another important domain is the networking of informatics systems and their communication by means of technical protocols. Basic principles of communication on the Internet should be explored 'unplugged' using simple playful means, since a basic understanding of how the Internet works is an important part of computer science education. Suitable examples include role-plays on sending mail via routers, calling up and forwarding entries on web pages to demonstrate the client-server principle or, as an example of the layer protocol of the Internet, telephone communication between people in different languages with the help of an interpreter.

**Computer Science and Society (C5)**

Even though this very important area of computer science education is not always easy to understand for children in learning scenarios, early childhood educators and primary school teachers should acquire basic knowledge about these connections.

The ubiquity of information technology has led to an intensive discussion in computer science about the social significance of computer systems and their design for the individual and society. The terms 'information society' or 'ubiquitous computing' are important concepts and indicators for this discussion. Applications and effects of informatics systems on society, as well as fundamental questions about the design and responsibility of information technology are also seen as an important part of general education and are therefore included in the computer science curricula of all educational levels from junior secondary onwards. However, since these are usually very complex contexts that are difficult for laypersons to understand, such topics are often only treated in an exemplary and very basic manner even at these school education levels, without diving too deeply into the informatics aspects of the problems. For the target group of children at the child-care centre and primary school level, the cognitive development and the ability to evaluate ethical and moral issues pose a hurdle when designing learning scenarios with such topics. If mobile devices with apps are used in the children's family environment or at school, the disclosure of personal data could be addressed in a child-friendly way in this context.

This is also an important topic for early childhood educators and primary school teachers, whereby one could, for example, analyse the usage profiles and data protection settings of frequently used apps or programmes (e.g., Facebook, WhatsApp). In this way, the basics of data and copyright protection could be made tangible for this target group by means of a specific example. In any case, it should be made clear which data traces are left behind when surfing the Internet and how these can be used for various evaluations that do not serve the original purpose. This can be done, for example, via settings for cookies in frequently used browsers or in an action-oriented and 'unplugged' way with a simulation game on data protection (Medienwissenschaft Universität Bayreuth, 2014).

## 4.3.2 Process Domains

As shown above in sections 2.5.1 and 3.3, the description of expected competencies in learning processes of computer science education also includes the representation of processes that are located in the respective Content Domains. It already became clear in the description of the examples in the Content Domains that these cannot be presented without corresponding processes that are carried out by the learners in the respective application contexts. Following the GI educational standards for the junior and senior secondary levels, as well as for the primary level, corresponding Process Domains should also be formulated in the expected competencies and objectives for teachers.

**Interacting and Exploring (P0)**

The importance of this Process Domain for children's computer science education has already been presented in detail in Chapter 3. Early childhood educators and primary school teachers should know the levels of reasoning for the introduction of this Process Domain in computer science education with children and be able to organise age-appropriate learning scenarios for computer science education for the children on the basis of this knowledge.

It is important that early childhood educators and primary school teachers, on the one hand, give the children enough room for their own explorations, but, on the other hand, also systematise the children's experiences in a suitable way so that they can build up a mental model for handling, using and designing informatics systems. Early childhood educators and primary school teachers should take into account the following perspectives when organising learning scenarios with regard to the Process Domain 'Interacting and Exploring':

*Learning and Developmental Psychology Perspective*

In the first years of their life, children often learn by trying out and exploring objects in their living environment. Therefore, it makes sense to also use this form of learning in early computer science education. By observing and trying out suitable computer systems, children can learn about concepts of action and design by dealing with them. Based on the observed functions, children can make assumptions

about their inner structure. Early childhood educators and primary school teachers can then take up these experiences and assumptions of the children through the appropriate design of learning scenarios, systematise and generalise them with the children, in order to gradually open up a more differentiated picture of informatics systems for the children.

*International Perspective*

As has already been shown for the goals at the level of the children on the basis of numerous practical examples from projects in various countries, it is common practice internationally to let children in primary school and early childhood education explore informatics systems through play and to use them in various ways (cf. Chapter 3). Early childhood educators and primary school teachers should be familiar with some of the international concepts of computer science education for interacting with and exploring informatics systems. These concepts can also be used in computer science education with regard to their exploratory and systems research elements. For example, Resnick's methodological approach of 'Tinkering', which refers to action-oriented approaches of Dewey (1938), Fröbel (1826) and Papert (1993), can be practically implemented by means of the process dimension of 'Interacting and Exploring'.

> "Froebel filled his kindergarten with physical objects (such as blocks, beads, and tiles) that children could use for designing, creating, and making. These objects became known as Froebel's Gifts. Froebel carefully designed his Gifts so that children, as they played and constructed with the Gifts, would learn about common patterns and forms in nature" (Resnick, 2013, p. 50).

In the age of digital artefacts, Resnick's concept is extended in the sense of Papert to digital micro-worlds in the computer (e.g., Scratch programs) and digital objects in the real world (e.g., Mindstorms). This concept, applied in learning scenarios with children, is also viable for the computer science training of early childhood educators and primary school teachers.

> "We see tinkering as a valid and valuable style of working, characterised by a playful, exploratory, iterative style of engaging with a problem or project. When people are tinkering, they are constantly trying out ideas, making adjustments and refinements, then experimenting with new possibilities, over and over and over" (Resnick, 2013, p. 164).

*Subject-didactic Perspective*
In computer science didactics, great importance is attached to interaction with and exploration of informatics systems. This ties in with findings in learning psychology, such as the Cognitive Flexibility Theory (Spiro, Feltovich, Jacobson & Coulson, 1991), according to which an object area should be viewed from different conceptual perspectives at different points in time in order to gain a deeper understanding of it. In this sense, a playful, explorative exposure to informatics systems can be a starting point for later reflections and furthermore promote a self-confident and reflective approach to informatics systems in terms of self-efficacy concepts. This could provide a basis for sustainable computer science learning processes.

In the subject-didactic competence research, theoretically and empirically supported competence models for understanding and modelling informatics systems suggest that the exploration and application of informatics systems is an essential facet of competence for understanding informatics systems (e.g. Magenheim et al., 2010). Various subject-didactic concepts have so far tried to justify this approach to computer science education with different arguments. Proponents of the application-oriented approach in computer science didactics, Körber & Peters (1988) for example, emphasise that informatics systems should always be considered in an application context and that the use and testing of an informatics system represents an essential phase of computer science-related learning processes.

The user-oriented didactic approach (e.g. Forneck, 1992) goes one step further and reduces the constructive development of informatics systems to a minimum in favour of the almost exclusive use and evaluation of informatics systems. In the system-oriented approach, following the Cognitive Flexibility Theory, a multifaceted approach to informatics systems is recommended by using the method of deconstruction to explore informatics systems and their functions in a first exploration phase in order to be able to draw conclusions about their inner structure later on (Magenheim, 2008).

The subject didactic approach to duality reconstruction aims at picking up on learners' everyday experiences with informatics systems (e.g. mobile phones, digital media recorders, standard software) and exploring the functioning of digital artefacts. In a process of didactic reconstruction, the duality of structure and function of digital artefacts is to be developed, whereby the function refers to the purpose and use of the artefact in everyday life, the structure to its internal construction (Schulte, 2009).

The discovery and understanding of computer science concepts through exploring the functioning of informatics systems and the interplay of behaviour and structure, of external and internal view via the gradual change of the learning scenario with the informatics system from the 'black box' (only external view is accessible) to the 'white box' (internal structure of the informatics system, e.g. source code is accessible to learners) is justified in another approach by Stechert (2009).

Empirically and theoretically based competence models are also discussed in subject didactic competence research, which regard the ability to explore and apply informatics systems as an important facet of competence for understanding informatics systems (e.g. Magenheim, 2008).

Finally, the methodology of experimentation can be used for exploring informatics systems and digital artefacts. In this context, it has been shown that through purposeful interaction, it is possible to draw conclusions about internal structures on the basis of the system's feedback (Schulte, 2012).

As a consequence of these subject didactic arguments and empirical findings, for the goals at the level of the early childhood educators and primary school teachers, in addition to an 'unplugged' access to concepts of computer science, a handling of at least those informatics systems should be practised that can be used in learning scenarios with children. Furthermore, early childhood educators and primary school teachers should know some of the essential subject-didactic arguments for interacting with and exploring informatics systems and thus be able to organise corresponding learning scenarios in a sound manner.

*Computer Science Perspective*
Experts and users exploring and interacting with existing informatics systems is an important phase in the further and new development of informatics systems and their implementation in practice. In this way, from the basic, externally recognisable functioning of informatics systems, their design principles can be identified and new system components can be constructed or the system as a whole can be further developed (see, for example, Brandt-Pook & Kollmeier, 2008). Moreover, informatics systems and software should be designed in such a way that the functioning of the systems is largely self-explanatory even for laypersons and can be easily configured for their specific purposes. Computer science education should therefore also enable laypersons to specify and competently use informatics systems. In this way, computer science education makes an important contribution to general education.

For the goals for early childhood educators and primary school teachers, one can formulate from this perspective that they should also know the subject-related computer science arguments for the use and exploration of informatics systems and they should be able to implement them in practice in their own work context or by means of a small project. This could be done, for example, with a very simple game programmed in Scratch. The functionality of such a game could be explored, a possible functional extension from a child's point of view could be anticipated and then implemented through a slight modification of the source code.

*General Education Perspective*

Early childhood educators and primary school teachers should be aware of the general education aspects of children's exploration and use of informatics systems mentioned above and support them in suitable learning scenarios. Through engagement with informatics systems and appropriate facilitation by early childhood educators and primary school teachers, children can learn to successively build a more sophisticated mental model of informatics systems. In this way, they are taught to use informatics systems competently in the future and to evaluate technological change in terms of its opportunities and risks. In order for this demanding long-term goal to succeed, early childhood educators and primary school teachers should be able to strengthen children's self-efficacy and motivation when dealing with age-appropriate informatics systems, as well as to promote their ability to reflect with regard to the use of informatics systems. Corresponding competencies in the children, but also in the early childhood educators and primary school teachers, require interactive handling of such systems, which can be implemented with unplugged methods not only in learning scenarios.

**Modelling and Implementing (P1)**

Modelling of an informatics system and its (software-) technical implementation with the help of a programming language are important tasks in computer science and represent essential steps in the design of an informatics system. For the target group of early childhood educators and primary school teachers, this cannot be about getting to know different procedural models of software development. However, some important phases, methods and processes used in them should be identified, since these are also of great importance, for example, for the development of learning and game software. After a phase of requirements analysis on the software, the phases of modelling and implementation should follow. Finally, the resulting software should be tested and evaluated. Here, a connection could also be made to a specific discovery and research circle for informatics education (see Section 1.5).

One possible approach at this point would be the 'project-like' development of a small game using the Scratch programming language. Numerous examples of such a project can be found at https://scratch.mit.edu. In the 'project', various important phases could be played through. Such a concept would in particular also involve Process Domains P2-P5.

In the phase in which the requirements for software are determined, which precedes the actual modelling, it can be cooperatively determined which functions the (game) software to be developed should fulfil. In the modelling phase, essential components and parameters of the informatics system, as well as the relationships between them are then defined at the level of formal description. Also, the appearance of the user interface is fundamentally determined in this phase.

For early childhood educators and primary school teachers, a playful unplugged approach can also be used for this topic by modelling parts of the system using CRC cards (Ambler, 1998) or with a role play (object play) (Börstler & Schulte, 2005). However, these methods, which belong to object-oriented modelling, should not lead to the use of an object-oriented programming language in a possible later implementation. Neither can it be about working out different programming paradigms with this target group, nor about getting to know different programming languages. Even a visual programming environment with a rudimentary object-oriented concept like Alice would probably confront the target group with unnecessary learning barriers, as corresponding experiences with junior secondary level learners have shown (Dohmen, Magenheim & Engbring, 2009). In order to avoid medial gaps, early childhood educators and primary school teachers should basically stick to the programming language, which can also be used in learning scenarios with children if necessary (e.g., Scratch or a related dialect). Thus, in modelling, small problems can first be broken down into even smaller tasks and structured if necessary (see P3). Then, the algorithms determined in this way are presented as instructions for action (simple diagram) and implemented in a programming language suitable for children, e.g., Scratch. This can lead to visual solutions on the screen (e.g., computer game) or to moving objects in the real world (robot).

**Reasoning and Evaluating (P2)**

Decisions about the design of an informatics system must be made especially during the cooperative determination of system requirements and in the modelling phase. The same applies when a (self-developed) informatics system is tested and evaluated in its functioning and with regard to the intended problem solution. In both cases, criteria should be established according to which the computer science issues can be evaluated. When determining the system requirements, the functional scope (e.g., of a game) can be defined and an estimate of the effort required for realisation can be provided. When evaluating the finished software, e.g., ergonomic criteria of the user interface (e.g., Herczeg, 2009) and the usefulness of the system functions in a given social situation can be assessed (e.g., the possibility of playing a game with a group). In communication with other users and evaluators, modelling decisions can be justified and evaluations can be supported with arguments. In learning scenarios with informatics systems, early childhood educators and primary school teachers should be able to plan and evaluate exploration or design processes with informatics systems or action strategies in unplugged concepts cooperatively with the children.

## Structuring and Interrelating (P3)

Structuring and interrelating may be necessary processes in different Content Domains of computer science. As already mentioned in P2, modelling of an informatics system may require a structured break down into individual components. Sensible structuring is also required when selecting suitable data structures. Networking of system components

can occur in many different ways within an informatics system and can also include external communication with other systems, especially in informatics systems that are connected via local networks or the Internet. This concept can be illustrated by means of small robots with actuators and sensors that react to 'impulses' from their technical environment. In computer science learning scenarios, whether plugged or unplugged, early childhood educators and primary school teachers should illustrate the structuring of a problem area as a prerequisite for a solution, e.g., when a robot needs to know different directions of movement (left, right, straight ahead) in order to move in a maze. The networking of an informatics system can be illustrated by the playful transmission of an 'email'.

## Communicating and Cooperating (P4)

Computer science-based problems from different Content Domains are often not solved individually but cooperatively in a team. This requires general cooperative and communicative skills, which the team members practise at a formal level. In terms of content, professional skills to communicate the computer science material using computer science terminology are expected to contribute cooperatively to the solution of computer science problems. This communication and cooperation can be done "unplugged" using various non-digital materials as tools and in the medium of interpersonal communication. On the other hand, communication and cooperation can also be organised with the help of an informatics system. For this purpose, synchronous (e.g., chat, instant messaging...) and asynchronous communication possibilities (e.g., e-mail) of informatics systems can be used to exchange information and to organise collaboration. Electronic (web-based) platforms can also be used to exchange and archive documents. Early childhood educators and primary school teachers should be familiar with these computer science-related forms of communication and cooperation, and be able to apply them and use them in learning scenarios with children in a target group-oriented way. In doing so, they should pay attention to the correct use of technical terms and jargon in order to properly preconfigure the children's ideas when building mental models of informatics systems.

*Representing and Interpreting (P5)*
The knowledge acquired individually or cooperatively in the different computer science Content Domains must be communicated in a suitable way in order to receive and interpret it individually or discuss it with others. Various symbolic, graphic or pictorial forms of representation are suitable for this. The selection of suitable forms of representation to illustrate and interpret facts of computer science is thus an important Process Domain to describe computer science competencies. On the one hand, early childhood educators and primary school teachers should be able to select age-appropriate forms of representation that they can also use in learning scenarios with children. On the other hand, they should know and partly use forms of representation, which are suitable to visualise somewhat more complex issues in computer science (e.g., knowledge networks, flow charts, structure diagrams, graphs).

### 4.3.3 Contextualised competence expectations

In the following section, the combinations of individual Content and Process Domains considered to be particularly relevant for contextualised, subject-specific competencies that are expected from early childhood educators and primary school teachers are presented in a concise form.

In principle, the same argumentation applies here already at the level of the goal dimensions for children (cf. Section 3.4). In principle, each of the Content Domains can be combined with each of the Process Domains. Thus, expected competencies can be formulated that relate to each of these combinations. Suitable examples of learning scenarios can also be found for each of these combinations, but their presentation would go beyond the scope of this report. Moreover, it seems hardly feasible to develop a curriculum for our target group that covers all these combinations in a realistic time. But this is not necessary anyway. Many of the examples mentioned in the text above often refer to several Content and Process Domains, so that a new approach (learning scenario) is not necessary for each combination.

In a curriculum for early childhood educators and primary school teachers, it should be ensured that as many Content and Process Domains as possible are addressed at least once in one of the learning scenarios. Specific criteria apply to the selection and prioritisation of the subject-specific competencies expected of early childhood educators and primary school teachers. In this respect, reference can be made to the argumentation in Section 3.4. However, early childhood educators and primary school teachers should acquire basic subject-related competencies in the selected combinations of Content and Process Domains (C/P), whereas children, depending on their age group, will certainly only be able to acquire initial subject-related facets of competence. It is therefore proposed to select the following combinations of domains for early childhood educators and primary school teachers:

- Modelling and Implementing (P1) of Algorithms and Programmes (C2)
- Interacting with and Exploring (P0) Informatics Systems (C4)
- Representing and Interpreting (P5) Information and Data (C1)
- Reflecting on and Evaluating (P2) the interrelationship of Computer Science and Society (C5)

The selection is based on the guiding criteria already applied in Section 3.4 and Chapter 5 for the selection and prioritisation of competencies and examples to be taught:

- according to subject-specific and subject-didactic significance
- possibility of realisation under aspects of learning and developmental psychology
- relation to the children's everyday situation
- ability to motivate the learners
- arouse the children's interest in computer science and contribute to their general education
- possibility of also teaching interdisciplinary basic skills
- relation to didactic concepts of the primary school or child-care centre
- possibility of orienting towards the practice of concepts that have already been realised

Since each of the above four combinations has to be contextualised on the basis of specific practical examples, one example usually involves several combinations of Content and Process Domains, with one or two of the combinations forming a focal point, while other combinations are not so strongly addressed in the given example. Chapter 5 therefore presents the examples selected for prioritisation using a "heat map", where the colours indicate how strongly a C/P combination and the associated competencies are affected in the respective example.

For early childhood educators and primary school teachers, the above prioritisation and the prioritisation for the target group of children results primarily in competence expectations in the competence fields marked in green (C/P).

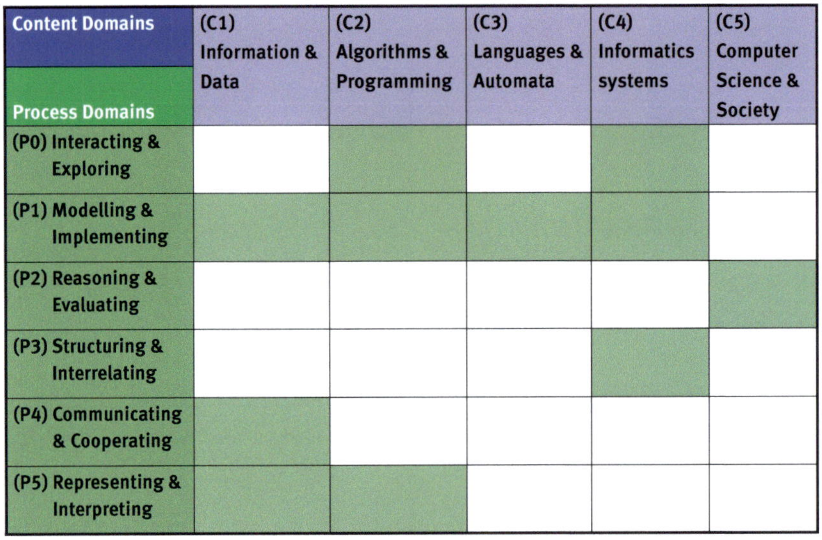

| Content Domains<br><br>Process Domains | (C1)<br>Information & Data | (C2)<br>Algorithms & Programming | (C3)<br>Languages & Automata | (C4)<br>Informatics systems | (C5)<br>Computer Science & Society |
|---|---|---|---|---|---|
| (P0) Interacting & Exploring | | | | | |
| (P1) Modelling & Implementing | | | | | |
| (P2) Reasoning & Evaluating | | | | | |
| (P3) Structuring & Interrelating | | | | | |
| (P4) Communicating & Cooperating | | | | | |
| (P5) Representing & Interpreting | | | | | |

*Figure 38. Competencies expected of early childhood educators and primary school teachers*

The examples presented in Chapter 5 illustrate how the concept could be practically implemented in learning scenarios appropriate to the target group. For the group of early childhood educators and primary school teachers, all the domains of competence mentioned in this chapter can be addressed on the basis of the learning scenarios planned for the children, if they also deal with the learning-psychological, subject-didactic and interdisciplinary competence domains in the learning group as an example outside the subject level.

## 4.4 Computer Science Didactic Competencies

As already described at the beginning of Chapter 4, we will be guided by Shulman's PCK concept and the corresponding implementations of mathematics didactics when presenting the subject didactic competencies, expected in the goal agreements from early childhood educators and primary school teachers. However, of central importance in this section are the concepts and models recently developed in subject didactic research in computer science. We will present them in somewhat more detail at this point and then specify them in the following subsections of the chapter in rather concise form with regard to the respective categories and the specific educational context in elementary and primary education.

Computer didactic competencies describe teachers' abilities to translate their subject-specific competencies into effective lesson design and learning scenarios for the learners in their target group. In teacher training, therefore, subject-specific, subject-didactic and practice-oriented training phases are closely interlinked. In most cases, subject-specific scientific phases precede subject-specific didactic phases, since the subject-specific didactic preparation of computer science content requires corresponding subject-specific knowledge. However, this is not to be understood as a linear and one-sided process. The situational learning of subject-related content in subject-related learning scenarios in lessons at the elementary and primary level can also contribute to the acquisition of subject-related and subject-didactic competence, e.g., for early childhood educators and primary school teachers. This must be kept in mind when organising learning scenarios for this target group.

In this context, Shinners-Kennedy and Fincher point out that teachers' professional reflection on an important subject area in computer science technology can change their subjective views of this and related subject areas (Shinners-Kennedy & Fincher, 2013). They illustrate this by looking at the 'Big Ideas' of computer science, using Meyer and Land's 'Threshold Concepts' (Meyer & Land, 2005):

> "It represents a transformed way of understanding, or interpreting, or viewing something without which the learner cannot progress. As a consequence of comprehending a threshold concept there may thus be a transformed internal view of subject matter, subject landscape, or even world view" (Meyer, Land 2003, p.1).

With his concept of the fundamental ideas of computer science, Schwill has provided criteria for identifying such 'Big Ideas', which can also serve as categories for selecting suitable learning content: Horizontal criterion (subject content can be explored at different cognitive levels), vertical criterion (subject content is relevant in many sub-disciplines of computer science), time criterion (subject content is relevant to computer science over a longer period of time), meaning criterion (subject content can be explored through the experience of the target group in their everyday world) and target criterion (subject content opens up a relationship to current research questions in computer science) (Schwill, 1993). For learning scenarios with children in primary schools and child-care centres, but also for early childhood educators and primary school teachers, content should be selected primarily with regard to the meaning criterion with its reference to everyday life. Furthermore, for this target group, in addition to the subject-specific significance of the topic, its contribution to general education should also be taken into account. Criteria from Klafki and Bussmann and Heymann can be used to assess this aspect (Bussmann & Heymann, 1987; Klafki, 1993). Klafki proposes the following criteria, among others: (meaningful) education for all, concerning all human abilities, concerning a key problem typical of an epoch, problem-oriented, combining subject-related and social learning. Bussmann/Heymann name the following as criteria for general education: Preparation for future life, foundation of cultural coherence, world orientation, instruction in the critical use of reason, development of a willingness to take responsibility, practice in understanding and cooperation, and strengthening of the learners' identity.

Buchholz, Saeli & Schulte, following Shinners-Kennedy and Fincher, as well as Saeli, each propose a PCK model for the acquisition of competencies by computer science teachers, specified for a particular content (Buchholz, Saeli & Schulte, 2013; Saeli, 2012; Shinners-Kennedy & Fincher, 2013). While Saeli uses the example of programming to show how teachers can acquire pedagogical subject competencies, Buchholz, Saeli and Schulte choose a more general approach and use the empirical instrument CoRe (Content Representation) to identify the PCK of a basic idea. This instrument aims to use appropriate questions to both justify the selection of the corresponding content as a learning object and to describe the way in which it can be taught in a learning scenario.

"CoRe involves the following series of questions:

1. What do you intend the students to learn about this Big Idea?
2. Why is it important for the students to know this Big Idea?
3. What else do you know about this Big Idea (and you don't intend students to know yet)?
4. What are the difficulties/ limitations connected with the teaching of this Big Idea?
5. Which knowledge about students' thinking influences your teaching of this Big Idea?
6. Which factors influence your teaching of this Big Idea?
7. What are your teaching methods (any particular reasons for using these to engage with this Big Idea)?
8. What are your specific ways of assessing students' understanding or confusion around this Big Idea?" (Buchholz et al., 2013, p. 10)

The questions related to a specific computer science topic also touch on important pedagogical fields of action in teaching practice, which need to be reflected upon in the sense of PCK-related competence acquisition. Through the interlinking of theory and practice and the gathering of practical teaching experience, student teachers of computer science can acquire in-depth PCK in a cyclical process of reflected practice on the basis of subject-specific and subject-didactic theory, which goes beyond purely superficial formal definitions of the media and methodological design of learning scenarios. Based on these theoretical considerations and on empirical data obtained in a computer science teacher training programme, the authors propose a two-dimensional PCK development model.

One dimension refers to the didactic-methodological decision-making level, which can be tapped with the questions cited above, and is grouped according to three sub-areas:

- teaching (what? why? depth of content?)
- learning (learners' prior knowledge, assessment methods)
- other factors (teaching material, institutional conditions, content-related methodological arrangement)

The second dimension represents a grading of the skills in the different areas of the first dimension on 3 levels, whereby a high level of reflection is expected at level 3 in each case, taking into account subject-specific and subject-didactic concepts and references to practice (Buchholz et al., 2013, p. 15).

| Level | Teaching nexus | | | Learning nexus | | | Other | |
|---|---|---|---|---|---|---|---|---|
| | Q1 (what) | Q2 (why) | Q3 (SMK, reduc.) | Q4 (difficulties) | Q5 (prior knowledge) | Q8 (assessment) | Q6 (forces) | Q7 (methods) |
| 1 | topic is named | goal is named but not justified | (nearly) no knowledge about the topic | teacher centric: tries to cope with the topic herself | mentions part of the content | teacher just knows (observes) | focus on organizational issues ans teching material; vague impression that pupils' perspective is important | few methods |
| 2 | topic and CS connected | NOT CLEAR [not: names more than one goal!] | deeper content knowledge, but no justification what to leave out | teacher centric: how to reduce/ reconstruct | NOT CLEAR | teacher inquires | NOT CLEAR | more methods, and justification for Content |
| 3 | connected to every day life | goal is justified; connection between CS and real world is made | focus on specific parts of the deep content knowledge, no justification what to leave out | learner centered; knows specific, content-related learning obstacles | NOT CLEAR | teacher has methods/ aspects of inquiring | NOT CLEAR | broad knowledge of methods; use of methods is partially justified |

*Figure 39. Preliminary competence model (Buchholz et al., 2013, p. 15)*

The competence model developed in the Federal Ministry of Education and Research (BMBF) project KUI "Kompetenzen für den Informatikunterricht" (Competencies for Computer Science Teaching) offers a further point of reference for the presentation of the computer science didactic competencies of teachers (Hubwieser et al., 2013). Based on a category system derived from the 'Darmstadt Model' (Hubwieser et al., 2011), which was developed as a computer science-related advancement of the learning theory models for lesson planning ('Berlin Model', Heimann, Otto & Schulz, 1979; Hamburg Model, Schulz, 1997), a competence structure model was first derived normatively-deductively. This was done on the basis of a qualitative content analysis (Mayring, 2010) of training curricula of computer science-teaching degree programmes, primarily at German but also foreign universities, and of relevant subject didactic literature. The competence structure model developed in this way was then empirically differentiated and tested with the help of expert interviews using the 'critical incident method' (see Hettlage & Steinlin, 2006). In a further step, a measuring instrument was developed and used which, with data from different target groups (experienced computer science teachers, student trainees, computer science-teaching students), on the one hand, provided indications of the corresponding expertise of the groups surveyed. On the other hand, it also offered the opportunity to check the validity of the underlying competence model (Hubwieser et al., 2013). Although the competence model and the associated measurement instrument were again primarily developed for computer science teachers at secondary schools, the categories contained in the competence structure model can also be adapted contextually and provide important information for the competencies expected of early childhood educators and primary school teachers in elementary and primary schools (both in and out of school). As a result of the normative-deductive analyses, a two-dimensional competence structure model emerged, which was later reduced somewhat for reasons of practicability during the empirical refinement (cf. Fig. 40).

- *Dimension 1:* Fields of Pedagogical Operation (FPO), including sub-categories
  - FPO 1: Planning and design of learning situations
  - FPO 2: Reacting on student's demands during teaching process
  - FPO 3: Evaluation of teaching processes
- *Dimension 2:* Aspects of Teaching and Learning (ATL) with 15 sub-categories, which in turn are divided into 5 groups:
  - Group 1: Subject and curriculum related issues
  - Group 2: Teaching methods and use of media
  - Group 3: Learner related issues
  - Group 4: Teacher related issues
  - Group 5: Issues of the educational system

In detail, the model for ATL dimensions looks as follows:

|  | FPO 1 | FPO 2 | FPO 3 |
|---|---|---|---|
| ATL 1 |  |  |  |
| ATL 2 |  |  |  |
| … | … | … | … |
| ATL 15 |  |  |  |

| Cat. Nr. | Field descriptor | Subcategories |
|---|---|---|
| FPO 1 | Planning and design of learning situations | – Time planning (Time allocation),<br>– Explanation of the planning: subject specific consistency, reasonability of the approach, psychological argumentation<br>– Granularity: long term lesson planning, planning the entire curriculum, planning a lesson |
| FPO 2 | Reacting on student's demands during teaching processes | – Reacting based on understanding: flexible use of connected knowledge in critical situations, responding to students appropriately, responding flexibly<br>– Mastering complexity<br>– Keeping compliant with planning |
| FPO 3 | Evaluation of teaching processes | – Techniques,<br>– Criteria<br>– Derive consequences |

| Cat. Nr. | Category | Subcategories |
|---|---|---|
| **Group 1: Subject and Curriculum related Issues** | | |
| ATL 1 | Learning content | – Multiple representations<br>– Category systems for learning content<br>– Specific school-related content<br>– Selection and justification of learning content<br>– Didactical (re-) construction of subject-matter knowledge |
| ATL 2 | Subject | – Relations to other subjects<br>– Definition of computer science education<br>– History of computer science education<br>– Relationship of the subject to the scientific discipline<br>– Objectives of the subject<br>– Legitimacy and relevance of the subject |
| ATL 3 | Curricula and standards | – Curriculum development<br>– Relation to other subjects<br>– Approach and structure of the curriculum<br>– Selection and commitment<br>– Actual examples of curricula |

| ATL 4 | Objectives of lessons | – Focus on education standards<br>– Competencies<br>– Learning objectives |
|---|---|---|
| ATL 5 | Extracurricular activities | – External collaboration<br>– Contests |
| ATL 6 | Science | – Subject discipline<br>– Computer science education as a scientific discipline<br>– Relationship between teaching of the subject and the scientific discipline |

| Group 2: Teaching methods and use of Media | | |
|---|---|---|
| ATL 7 | Teaching Methods | – Organizational arrangements<br>– Methodological principles<br>– Subject-specific teaching methods |
| ATL 8 | Subject-specific teaching concepts | – Introductory lessons<br>– Programming classes<br>– Historical approach |
| ATL 9 | Specific teaching elements | – Lab-based teaching<br>– Experiments<br>– Tasks and assignments |
| ATL 10 | Media and educational material | – Application of hardware and software<br>– Textbooks<br>– Unplugged media |

| Group 3: Learner related Issues | | |
|---|---|---|
| ATL 11 | Heterogeneity in the context of subject-specific learning | – Age<br>– Gender<br>– Ethnical background<br>– Family socialization<br>– Disabilities |
| ATL 12 | Student cognition | – General subject-related cognitive aspects<br>– Individual learning Diagnostics, performance evaluation and assessment<br>– Cognitive activation |

| Group 4: Teacher related Issues | | |
|---|---|---|
| ATL 13 | Teachers' perspective | – Collaboration<br>– Core tasks<br>– Qualification<br>– Motivation<br>– In-service training<br>– Teaching experience |

| Group 5: Issues of the Educational System | | |
|---|---|---|
| ATL 14 | School development | – Policies<br>– Quality management<br>– School profile |
| ATL 15 | Educational system | – School type<br>– Enrollment<br>– Organizational aspects of subject |

*Figure 40.* Competence structure model for computer science teachers (Hubwieser et al., 2013)

In the empirically based further development of the competence structure model, it became apparent that, similar to the model by Buchholz, Saeli and Schulte described above, the ability of teachers to combine subject-specific and subject-didactic knowledge and to translate this into practical, learner-related teaching operation is an important distinguishing criterion for the degree of computer science didactic competence acquired (Buchholz et al., 2013).

In the following, we will briefly specify the findings from the presented competence models and publications on the subject didactic competencies of computer science teachers with regard to some relevant competence aspects for early childhood educators and primary school teachers. In doing so, we will also use adapted indicators from the KUI measurement instrument as well as adapted descriptions of competence from the subject-specific KMK descriptions of competence (Kultusministerkonferenz, 2015). Due to the partial interdependencies, the descriptions of the competence categories are not always completely disjunct. On the subject-specific level, we refer to the competence expectations described in Section 4.4. With regard to the organisational design of computer science-related learning scenarios for early childhood educators and primary school teachers, according to the findings of the aforementioned studies, a cyclical design of courses should be considered that …

- closely aligns the acquisition of computer science competencies with the learning scenarios for the children,
- combines subject-specific and subject-didactic problems with each other, and
- includes the practical implementation of previously jointly developed learning concepts and learning scenarios for children in a practical trial phase and their subsequent reflection.

### 4.4.1 Basic Computer Science-Didactic Competencies

This section describes the basic computer science-didactic competencies that the early childhood educators and primary school teachers should have in order to be able to successfully design target group-related learning and educational processes of computer science education in primary schools and child-care centres (see also Kultusministerkonferenz, 2015, p. 32). Early childhood educators and primary school teachers should have sound didactic and pedagogical-psychological knowledge of computer science education for children at the primary and pre-school level and be able to organise learning scenarios for computer science education that are appropriate for the target group, taking into account the developmental and learning-psychological abilities of the children (cf. also Section 2.1).

Early childhood educators and primary school teachers ...

- are able to interpret and evaluate educational standards, recommendations and curricula for computer science education in primary schools and child-care centres and to plan and implement their learning scenarios in line with these recommendations;

- can relate these recommendations in an appropriate way to requirements from orientation plans at the elementary level in order to derive criteria for the design of learning scenarios;

- are familiar with important subject didactic concepts of 'computer science education' (e.g., basic ideas of computer science, 'CS unplugged', 'computer science in context') and their importance for general education and are able to take them into account when planning their learning scenarios;

- are familiar with important forms of learning typical for computer science (e.g., explorative learning, application-oriented learning), as well as important phases of the development of small software projects and are able to take these into account when planning their learning scenarios;

- are able to relate and reflect on current basic subject-specific and subject-didactic developments in computer science at the primary level and to include them in the planning of learning processes for computer science lessons;

- are able to create initial basic ideas about the subject of computer science in the children through the conceptual interconnection of concepts of computer science;

- are familiar with important Content and Process Domains of computer science and are able to present these in target group-specific and everyday language in order to contribute to the development of computer science terms and concepts in the children;
- are able to identify computer science facts in different application references and factual contexts from the children's world of experience, to recognise their social significance at least to some extent and to use them for the design of learning scenarios of computer science education;
- are familiar with possibilities of illustrating basic computer science principles that appeal the visual, auditory and haptic perception of the children;
- are able to use their teaching experiences from the primary or elementary school level for computer science education and develop their own routines for pedagogical work in informatics learning scenarios;
- are able to motivate themselves and the children for computer science topics by creating authentic learning situations and contributing their interest in the subject of informatics.

### 4.4.2 Competence for Planning of Computer Science Learning Environments and Learning Situations

An essential element of the subject didactic competence of early childhood educators and primary school teachers is their ability to design suitable learning scenarios in relation to computer science education. The sub-competencies required for this are described in this section. The general pedagogical and subject-didactic competence for planning computer science learning scenarios is just as important as the ability to contextualise the planning for specific organisational and pedagogical framework conditions.

Early childhood educators and primary school teachers ...

- are familiar with the essential elements of computer science learning environments and use this knowledge for the targeted construction of learning scenarios through an appropriate selection of topics that are authentic for children, as well as adequate media and methods for their development;
- are able to assess their planning of computer science learning processes with children with regard to the necessary temporal aspects of learning phases and take into account the children's individual learning processes;

- are able to assess their planning of computer science learning processes on the basis of professional and learning-psychological criteria, especially with regard to the children's cognitive development;
- are able to select suitable learning contents for learning scenarios of computer science lessons and to justify the selection in terms of subject didactics ('Big Idea') and learning psychology;
- are able to define the competencies to be acquired by the children in computer science learning processes individually according to their performance on the basis of educational standards of computer science;
- are able to identify typical computer science pre-concepts, understanding and learning barriers of the children and to implement necessary formalisations and abstractions of computer science principles according to the children's cognitive abilities;
- are able to present tasks and learning contents of computer science lessons in a target group-oriented, child-friendly manner and in various simplified but correct forms, as well as to make abstract computer science terms (e.g., algorithm, date, information, variable…) tangible for the children through various examples and through playful and investigative occasions of exploration;
- are able to use motivating social forms typical for computer science lessons, such as group work on informatics systems, role plays or CS unplugged methods, in such a way that the children's competencies are promoted;
- are able to select software suitable for children, e.g., programming languages or modelling tools and simple age-appropriate informatics systems (see Chapter 2) for computer science education at the elementary and primary level and to integrate them sensibly into learning scenarios;
- are able to address gender-specific differences in early computer science education, especially in the selection of tasks, and to counteract the formation of stereotypes.

### 4.4.3 Action Competence in the Context of Informatics and Didactics

This competence domain describes the expected abilities of early childhood educators and primary school teachers to specifically implement their planning for computer science learning scenarios in the practice of primary school education and in learning processes in child-care centres and to steer these in a suitable manner. This includes the following competencies:

Early childhood educators and primary school teachers …

- are able to motivate children to learn computer science as well as to promote and evaluate individual learning progress within the framework of their pedagogical activities in computer science learning scenarios;

- know subject-specific intervention possibilities and are able to apply them in computer science learning scenarios according to the situation (e.g., in the case of computer science misconceptions, dealing with provisional and imprecise technical terms, reacting to children's thought constructions, heuristic assistance);

- recognise situations with computer science content in computer science education processes, take them up in an appropriate way if necessary and initiate further computer science-related actions and conversations in this way;

- are able to deal with learning difficulties of individual children also in situations of group work with informatics systems and to provide targeted individual support;

- are able to practically implement concepts for learning a simple programming language in computer science learning processes with a playful, also unplugged approach;

- are able to apply methods for activating children in computer science learning processes at the elementary and primary level in a creative and motivating way in order to control and promote the learning process;

- are able to identify different individual performance levels of the children in computer science education processes during practical pedagogical action and to react appropriately to this heterogeneity through action suitable for the situation (e.g., process aids, naturally differentiated play and exploration environments and learning arrangements);

- encourage the children in computer science learning processes to learn independently, to evaluate their achieved computer science solutions independently, and thus to promote their self-efficacy in computer science competence domains;

- promote the children's reflection on their own actions with computer systems through group discussions and ensure a cooperative determination of the learning success and positive feedback of the achieved results to the children.

### 4.4.4 Competence to Diagnose and Evaluate Computer Science Learning Situations

In addition to contextual didactic competencies in computer science, which describe the ability to act appropriately in education processes of computer science, early childhood educators and primary school teachers also need the competence to analyse learning situations in the subject with appropriate empirical methods in order to be able to draw conclusions for their own future pedagogical action in the learning group. Subject areas are indicators that concern the children's acquisition of competencies on different levels (subject-related, social, motivational and volitional competence structure elements; see Magenheim et al., 2010), as well as relevant criteria of the pedagogical action context related to computer science (cf. Section 4.4).

The pre-requisite for this competence is also a general diagnostic ability of the early childhood educators and primary school teachers. This also includes a general diagnostic ability in learning psychology with regard to the learning ability of the learning group to be supervised as a prerequisite for planning learning scenarios.

Expected competencies in this domain are:

Early childhood educators and primary school teachers ...

- know the basics of empirical competence assessment, performance diagnostics and performance evaluation in computer science education, are able to understand their results and interpret them with regard to the design of computer science education processes in their educational institution (e.g., intelligence and computer science performance tests);
- can analyse children with regard to computer science learning processes, their individual ways of thinking and ideas as well as with regard to their personal learning prerequisites, previous experiences and abilities with the help of computer science-related diagnostic instruments (e.g., 'Biber tests' http://www.informatik-biber.de);

- are able to create suitable assessment schemes for the computer science learning situations they organise, taking into account the methodological and media design used, and to assess the children's performance in these learning situations;
- are able to reflect on the computer science learning situations they have designed by means of empirically obtained data from their own learning group and, building on this, to revise their original planning for the learning scenario when repeating individual sequences;
- are able to motivate children for learning computer science, as well as to promote and assess individual learning progress of children on the basis of appropriate data;
- are able to observe the development of children in computer science education and diagnose any need for intervention with regard to computer science learning processes in the educational institution and in the family environment (e.g., in dealing with digital media).

### 4.4.5 Competence of Professional Pedagogical Communication with Parties Involved

Early childhood educators and primary school teachers at the elementary and primary level often carry out their work in close cooperation with their colleagues and with the children's family environment. In this pedagogical field of action, too, professional pedagogical communication competencies with all participants are necessary from the perspective of computer science education.

Early childhood educators and primary school teachers …

- are able to explain the learning scenarios and the intended educational goals of computer science comprehensibly to the children's parents and thus involve the children's family environment in the educational process in a motivating way;
- are able to organise learning processes related to computer science in their educational institution in cooperation with their colleagues and to communicate with them appropriately on a subject-related and subject-didactic level when designing learning scenarios and informatics lessons;
- are able to use forms of synchronous and asynchronous communication via digital media in addition to direct interpersonal communication in order to organise and coordinate learning processes related to computer science in

their educational institutions and to exchange documents via digital learning platforms;

- can use their knowledge of basic functionalities and pedagogical application possibilities of digital media at the elementary and primary level to advise colleagues and parents on the use of media by children in the pedagogical context.

## 4.5 Key Competencies for Dealing With Digital Media

As explained above, the subject-specific and subject-didactic competencies of early childhood educators and primary school teachers also include a basic knowledge of the functioning of digital media – as specific informatics systems – and their possible applications in the classroom and in learning scenarios related to computer science (see Schulte & Knobelsdorf, 2011; cf. also Chapter 1). On the one hand, this knowledge is technically specified with regard to its application in pedagogical contexts. On the other hand, the competencies acquired in learning contexts of computer science for understanding and using digital media can also be used by teachers in various other subject contexts and e.g., in counselling interviews with parents. With their concept of TPACK (Technological Pedagogical Content Knowledge), Mishra and Köhler have clearly worked out the connections between the subject-related pedagogical use of digital media and the subject-specific, subject-didactic, technical and pedagogical competence components (Mishra & Koehler, 2006).

According to this model, the subject-specific competencies and subject-didactic competencies (see above: e.g., programming with Scratch, use of tangibles such as Ozobot, MakeyMakey, Mindstorms, etc.) acquired by early childhood educators and primary school teachers, represent a basic subject-specific TPACK, which can, however, also be transferred to other pedagogical contexts. In this respect, this kind of computer science-related competence acquisition of early childhood educators and primary school teachers can also contribute to their digital media literacy and their competent handling of digital media in various contexts of use.

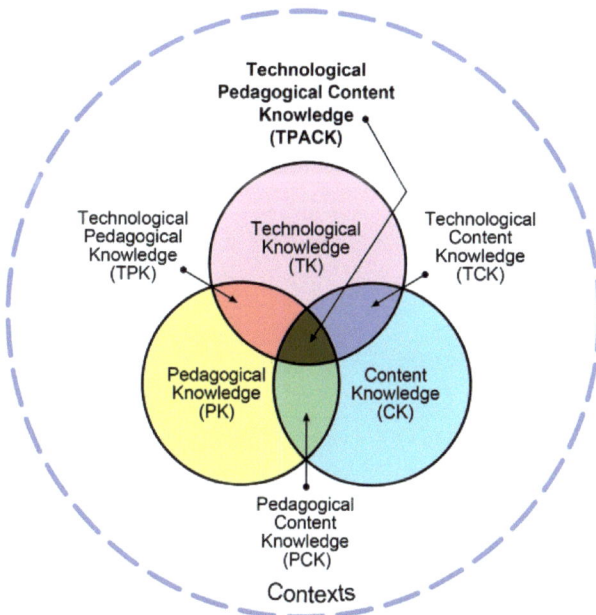

*Figure 41.* Structure of 'TPACK' (Mishra & Koehler, 2006, see also http://matt-koehler.com/tpack2/using-the-tpack-image/)

This is an important aspect of education. For example, the ISTE Teacher Standards require all teachers to have competencies in 'Design and develop digital age learning experiences and assessments', 'Model digital age working and learning', and 'Promote and model digital citizenship' (ISTE – The International Society for Technology in Education, 2008). Magenheim, Schulte and Scheel have presented a concept that identifies relevant computer science-content as important elements of digital media literacy for teachers of all subjects and thus clarifies the connection between 'digital media literacy' for teachers and their knowledge of fundamental concepts of computer science (Magenheim, Schulte & Scheel, 2002).

The importance of such competencies to understand and use digital media in educational activities has been presented in detail: with regard to the diverse possibilities of using digital media in schools, e.g., in Albers, Magenheim and Meister (2011), in Tillman, Fleischer and Hugger (2014) or Breiter, Welling and Stolpmann (2010) and with regard to the resulting expected competencies of teachers at the national level, for example, in the BMBF research project M3K on the media competence of teachers (Grafe & Breiter, 2014), in the KMK resolutions on media education (Kultusministerkonferenz, 2012) or internationally on the part of UNESCO (UNESCO, 2012) with its 'ICT-Literacy Concept for Teachers'.

In this sense, the goal dimensions described here for early childhood educators and primary school teachers can also make a specific contribution to digital media education for this target group.

Furthermore, in the sense of the three-dimensional OECD DESECO concept, educators can also be taught key qualifications for dealing with digital media in the information society: the ability to use digital media independently in different contexts of action, the ability to interact in heterogeneous groups and context-specific selection of appropriate digital tools (OECD, 2005).

## 4.6 Conclusion/Recommendations

The explanations in Chapter 4 have shown that adequate education and advanced training of early childhood educators and primary school teachers is an essential prerequisite for the successful implementation of computer science education in child-care centres and primary schools. Education and advanced training of early childhood educators and primary school teachers should be practice-oriented and geared to learning scenarios to be organised for the children (cf. examples and prioritisation proposals in Section 5.1). In this context, early childhood educators and primary school teachers should acquire basic subject-specific and subject-didactic competencies in the fields addressed here, which enable them to have sovereign pedagogical competence in organising and evaluating the computer science learning processes with the children. Furthermore, practice-based (practice-integrated) training of early childhood educators and primary school teachers should promote their motivation to design education processes, and their basically positive attitudes towards computer science and informatics systems should be fostered. This also includes the competent and critically reflective handling of digital media, especially in their function as learning media for computer science education. On this basis, early childhood educators and primary school teachers can then also acquire the necessary competence of professional communication with various groups of people involved in educational processes within their institution in order to successfully implement early computer science education in child-care centres and primary schools.

# 5 Examples of Prioritised Competence Domains for Computer Science Education

As already shown in sections 3.4 and 4.3, competencies can only be acquired in specific contextualised learning situations in which domain-specific action sequences are realised in selected Content Domains. By means of selected examples, this chapter shows how such learning scenarios of computer science education can be implemented successfully. The selection of the competence domains as a combination of Process Domains and Content Domains was justified in sections 3.4 and 4.3. It is based on the 10 criteria mentioned there, which are oriented towards subject-specific, subject-didactic, learning-psychological and general education aspects, among others. The examples of implementation were chosen in such a way that they cover the already prioritised fields of competence (C/P):

- Modelling and Implementing (P1) of Algorithms and Programmes (C2)
- Interacting with and Exploring (P0) Informatics Systems (C4)
- Representing and Interpreting (P5) Data and Information (C1)
- Modelling and Implementing (P1) of Languages and Automata (C3)

As shown above, the examples primarily refer to one or two fields of competence but usually also touch on several others. To illustrate this, a "heat map" of the competencies addressed is also provided for each example. Dark colours symbolise a strong representation of the competence in question in the respective example. Conversely, lighter colours indicate that the relevant competence is only marginally addressed in the example.

A summarising heat map is presented at the end of Chapter 5 (cf. Section 5.2) to clarify the recommended prioritisation of competencies as a whole. Independent of context, it illustrates which competencies (combinations of Process and Content Domains) are generally recommended for pedagogical implementation. For this purpose, only the primarily selected competence from each example is included in the summarising heat map. However, the Content Domain 'Computer Science & Society', which we, like many national and international curricula for junior and senior secondary sections, consider highly important is not a priority area for the target groups of this recommendation. Nevertheless, it should be integrated into the learning scenarios in a child-friendly way in individual examples, as shown below.

Since, from a subject-didactic point of view, the competencies to be acquired in the individual examples are independent in terms of content and early childhood educators and primary school teachers can design their complexity in adapted learning scenarios according to the cognitive abilities of the learning group, as well as to their needs and interests, it is difficult for the team of authors to assign the examples to age levels such as child-care centre or primary school. In this case, the practical implementation of the proposed examples would first have to be evaluated by means of empirically based accompanying research, in order to be able to make recommendations for grading on the basis of the experience gained.

## 5.1 Examples of Early Computer Science Education

### 5.1.1 Interacting With and Exploring Informatics Systems

**Prioritisation: C4** (Informatics Systems) **and P0** (Interacting and Exploring)
Also addressed: C2+P0 and C2, C4+P1

*Example*: **Exploration of an informatics system with LEGO WeDo**

**Executive summary**

Children use LEGO WeDo and instructions to build an informatics system, e.g., a robot or a controlled animal. Alternatively, they can use a system that has already been assembled.

This system consists of motors, sensors and actuators that can be programmed and controlled in combination with the corresponding software.

Children can explore the individual components of the system and their interaction, also in combination with programming. In this way, they learn about the components of the system and their interaction. In further steps, the system can be programmed and competencies in the field of algorithms and programming can be developed. Here, too, an explorative approach can be followed.

**Objectives**

By exploring the informatics system, children learn about the different components of the system and their functionalities. For example, the system consists of sensors and actuators. By exploring the system, children learn that, for example, actuators always react when the sensors pick up a signal (e.g., a noise or brightness). At this point, the IPO principle can also be addressed, for example.

Furthermore, insights into mechanical functionality vs. computer-science functionality can be gained, e.g., by replacing gears.

In a further step, programming can be added. Here, the children can explore and try out how, for example, motors are controlled or how actuators generally wait for inputs from a certain sensor.

## Heat map

| Content Domains<br><br>Process Domains | (C1)<br>Information & Data | (C2)<br>Algorithms & Programming | (C3)<br>Languages & Automata | (C4)<br>Informatics systems | (C5)<br>Computer Science & Society |
|---|---|---|---|---|---|
| (P0) Interacting & Exploring | | Exploring behavioural patterns, Conducting experiments | | Orienting and exploring of components and functionality | |
| (P1) Modelling & Implementing | | Systematically experimenting and developing algorithms | | Creating models of system functionality | |
| (P2) Reasoning & Evaluating | | | | | |
| (P3) Structuring & Interrelating | | | | | |
| (P4) Communicating & Cooperating | | | | Communicating about the discovered functionality | Discussing the impact of automation for society |
| (P5) Representing & Interpreting | | | | | |

## Description

LEGO WeDo is very well suited for exploration-based work. Learning theory focuses on constructionism, which was founded by Seymour Papert (1991).

Children investigate a frequently occurring real-life problem, e.g., the kicking range of a football player or sorting objects. To do this, they can first carry out numerous experiments and investigations that are either already carried out with the constructed system (see example of football player) or lead to the development of ideas for constructing their own system (see example of sorting). In a further step, they can modify the existing system or construct their own new system. In this way, the children learn about the construction and functioning (including a little insight into programming) of the system, similar to the idea of "tinkering".

*Possible examples:*

**Football player:** Here, for example, the kicking distance can be determined for a given ball. In further steps, the lever can be changed, for example, and the shooting distance can also be determined. The next step could be, e.g., to add a sensor that waits a certain time for a ball, or complement the system so that the player automatically returns to the starting position, etc.

**Sorting machine:** Learners solve a real problem, investigate how machines for sorting objects work and develop a sorting machine, check its functionality and experiment with different initial situations.

### Justification

By exploring and purposefully investigating an informatics system (both at the hardware and software level), children learn about and become enthusiastic about new informatics systems. They try out the systems and discover new phenomena. To better understand the behaviour, they create models to describe the behaviour. In further steps, children can participate in the design of informatics systems, e.g., by making adjustments to them. In this context, they also learn to deal with errors.

It can be assumed that these competencies are helpful in dealing with new informatics systems. In today's world, we are constantly confronted with new IT systems. By operating the systems, we construct mental models for these systems. Since the basic structure of informatics systems is always the same, it can be assumed that the acquired competencies contribute to more confident handling of informatics systems.

### 5.1.2 Programming and Algorithms

**Prioritisation:** C2 (Algorithms and Programming) **and** P1 (Modelling and Implementing)
Optionally addressed: P3 + P4

*Example A*: **Storytelling**

### Executive summary

Children create a short story or animation with a suitable programming tool. To do this, they look at templates, modify them and develop their own little stories on this basis.

They get to know the programming environment, a selection of basic programming instructions, and the structure of the examples and adapt them to their own ideas (in the literature, adapting foreign examples is also called 'remix').

### Objectives

Programming at this level: Expanding the expressive competence. Programming understood as: "define now, execute later/let others execute".

Expressive competence: The ability to express and share one's own ideas and thoughts with others.

Algorithms: Basic algorithmic building blocks: Instruction, branching, looping and possibly interaction, etc. (cf. e.g., Scratch overview page)

**Heat map**

| Content Domains / Process Domains | (C1) Information & Data | (C2) Algorithms & Programming | (C3) Languages & Automata | (C4) Informatics Systems | (C5) Computer Science & Society |
|---|---|---|---|---|---|
| (P0) Interacting & Exploring | | Adapting | Refining step by step | Exploring[40] | Communicating via programmes |
| (P1) Modelling & Implementing | | Planning a sequence | | | |
| (P2) Reasoning & Evaluating | | | | | |
| (P3) Structuring & Interrelating | | | | | |
| (P4) Communicating & Cooperating | | Communicating via programmes | | | Programming stories |
| (P5) Representing & Interpreting | | Storyboarding | | | |

**Description**

The idea of linking introductory programming with (animated) storytelling, probably originated in different places. The idea was developed as part of her PhD project by Caitlin Kelleher (2006), who was looking for ways to introduce girls to programming. She experimented with Alice (a 3D environment) and observed that children aged around 9-15 enjoyed using the system to create small stories, employing the important basic building blocks of programming. She formulates the advantages of the approach in her dissertation as:

> "Storytelling is a good context for middle school girls to learn about computer programming: 1. Given a little bit of time, most girls can come up with a story they would like to tell. Storytelling is, at its core, a form of communication which is an important activity to most middle school girls. 2. Stories are naturally sequential, allowing users to begin by creating sequences of instructions and gradually progress to more advanced programming concepts as they

---

40 Exploring the functions of the programming environment as well as, e.g., in troubleshooting, general exploration of how the computer works (for specialists: exploring the 'notional machine')

gain experience and confidence. 3. Stories are a form of self-expression and provide girls an opportunity to experiment with different roles, a central activity during adolescence. 4. Non-programming friends can readily understand and appreciate an animated story, which provides an opportunity for girls to get positive feedback from their friends" (Kelleher, 2006, p. 32).

Compared to working with a more traditional programming environment/approach (Kelleher, 2006, p. 38), the children did not learn more but spent more time on the activity of programming and expressed a stronger interest in continuing to programme with the system on their own and/or in further courses. Kelleher concludes:

> "Storytelling can provide a gentle, motivating introduction to programming concepts. Girls often begin by creating sequences of instructions and, as they gain confidence, create new scenes and new actions for their characters, tasks which often require more complex programming constructs. Girls' storyboards commonly included motivation to use methods, parameters, loops, and parallel execution" (Kelleher, 2006, p. 39).

The idea was implemented as so-called 'storyboarding', which can be assigned to the competence domain "Modelling":

> "In the first step, they (the children) wrote a single-paragraph description of the story they were planning to create. In the packet, I encouraged them to think of this paragraph as being similar to the description one might find on the back of the DVD box. In the second step, the worksheet directed girls to break their story into 3-5 separate scenes. For each scene, girls wrote a description of the setting, what happens during the scene (in 1-2 sentences), and the purpose for the scene (what the audience should learn from the scene). Finally, girls created a series of storyboard frames for each of their scenes. The worksheet provided 9 frames per scene and directed girls to both draw the frame and provide a short textual description of the action in that frame beneath it. In practice, most scenes contained 4 to 6 frames and accompanying textual descriptions" (Kelleher, 2006, p. 75).

Such storyboards do not necessarily have to be written but can also consist of small drawings:

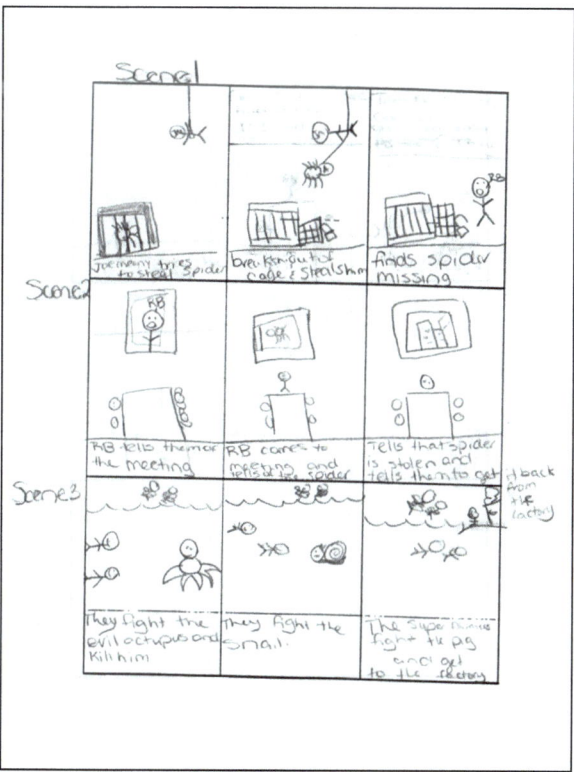

**Figure 42.** Example of a StoryBoard (Kelleher, 2006, p. 54)

Storyboards can be used to **model** processes and to convey the idea that the technique encourages people to adapt to new ideas as they progress.

When the potential of the storytelling approach became clear, Kelleher began to develop programming environments specifically suited to this approach. The environment from the PhD project was called StoryTellingAlice. She is now working on LookingGlass[41] with her own team, and is trying to put storytelling at the centre (Kelleher, 2015). It allows the simple representation of graphic characters, the interaction with these characters (cf. LookingGlass examples), easy use of graphic elements and controlling them with programme instructions, e.g., moving them or having characters speak, etc.

Since stories are mostly simple linear sequences, the choice of programming commands used is somewhat limited. Adams and Webster compare three differ-

---

41  https://lookingglass.wustl.edu

ent introductions to programming (young people from about 16 years) and find that the storytelling approach tends to use simpler algorithmic sequences:

> "Students creating games used the most variables, if statements, and loops. Students creating music videos used nearly as many loops as games, but far fewer variables and if statements. Students creating story-telling projects used the fewest loops, variables, and if statements, but made the most use of dialogue"[42] (Adams & Webster, 2012, p. 648)

According to Kelleher, however, there are still many possibilities:

> "Users of both Storytelling Alice and Generic Alice experimented with programming constructs beyond simple sequences. A majority of the participants in both groups used Do Togethers to have multiple animations occur simultaneously. 53% of the users of Storytelling Alice created a new method and used it in their program as opposed to 30% of the users of Generic Alice. 33% of the users of Generic Alice used loops as compared to 12% of the users of Storytelling Alice" (Kelleher, 2006, p. 184).

Scratch[43] is a similar environment. In a lecture,[44] Michael Resnick, one of the founders of Scratch, pointed out the many expressive possibilities of this programming environment. He was most touched by examples of greeting and congratulation cards that children had created for their grandparents, for example. These examples (which can be posted on the Scratch website) illustrate the importance of the programming environment for expanding expressive competence and also show that children have embraced the tool and activity of programming to the point where they can use it creatively for their immediate needs.

Similar experiences are repeatedly reported: Based on the exploration of similar examples, children quickly become active themselves and begin to change smaller things, first the colour and shape of the figure, if necessary, and eventually their behaviour to an increasingly greater extent. Children can largely (or even completely) control independent exploration and experimentation with individual

---

42 *Dialogue refers to method invocations that make the characters say something or think out loud*
43 https://scratch.mit.edu
44 https://www.ted.com/talks/mitch_resnick_let_s_teach_kids_to_code#

possibilities for change **(step-by-step refinement)** themselves. The environments also offer more and more support for gradual introduction[45].

In order to promote these own activities, children should first be given the opportunity to deal with programming environments (developed for children) in a playful way and without external requirements, in the spirit of the "Free Exploration and Experimentation" approach (FEE; see Köster, 2006). Observing the children during this initial orientation phase and the subsequent, often already somewhat more systematic exploration helps to recognise how the children proceed, which (often initially non-verbalised) questions they pursue and where they need support. Only when the children express the need for support or it becomes clear that they are not making progress on their own, can impulses or more open-ended tasks (such as "Tell your own story, draw your own moving picture" etc.) be helpful to strengthen creative expression and their own connection to the programming project.

An important feature of such tools and approaches is therefore the ease with which they can be inspired by other stories and use them as templates for their own stories (**adapting or remixing**).

If children are not doing this anyway, it is a good idea to encourage them to share ideas in between and summarise what possibilities and interesting programming commands they have already explored. Giving children the opportunity to present the structure of an example shows appreciation for what they have already achieved and can lead to shared further ideas[46]. Children can describe how they have created the flow of a story or how they intend to create it further. Storyboards can be drawn for this purpose, like in the case of films/comics (cf. example above, Figure 42).

In this way, children can also gain a first idea of what an algorithm is. Depending on the learning group, a reflection on the expression tools used with regard to designing a "flow" (as a term for algorithm) can take place in order to make clear that the story told represents an algorithm and that corresponding commands can be used: simple instructions and method invocations, possibly also loops and branching, etc.

**Justification (acquired competencies)**

Programming can and should be seen as a contribution to the development of expressive ability (Schulte 2013), in the context of **general education**: In this sense, a programme is an aesthetic artefact, a medium of expression and communica-

---

45  *ScratchJr (https://www.scratchjr.org/) is an example of such recent development. The programming environment is aimed at children aged 5-8.*
46  *There are materials on Google's storytelling approach that could be adapted for our target group if necessary: https://csfirst.withgoogle.com/c/cs-first/en/storytelling/materials.html.*

tion. Children can and should learn to use this form of expression and understand how it works. Programming exercises refer to the production of a mostly visual and animated, creative and artistic result and are mostly self-serving, similar to pictures painted on paper.

During the creation of the product (= programming) the goal may change, partly because new possibilities of expression are recognised, partly because new ideas emerge. This kind of programming is therefore self-directed (mostly within a framework) and thus contributes to **self-efficacy** in dealing with informatics systems and to the development of **interest** in programming and in the functioning of programmes and promotes the **motivation** to continue programming independently. A **life-world** and **everyday relevance** results from use in communication and design contexts in which small self-made aesthetic products play a role (e.g., greeting cards for certain occasions).

*Example B:* **Programming Unplugged**

**Executive summary**

There are numerous unplugged approaches for first experiences with programming. Many of these approaches focus on controlling a robot.

In some approaches, children themselves are the robots and are guided through a course by other children with specific commands. In other approaches, the robot characters are controlled, for example, over playing fields. The programming is done with cards or puzzle-like building blocks, for example.

**Objectives**

The aim is for children to playfully learn and understand the first steps of programming as an important concept in computer science through trial and error and experimentation. Thus, from the example of robots, they learn that you have to give them precise instructions in order to control them.

## Heat map

| Content Domains<br><br>Process Domains | (C1) Information & Data | (C2) Algorithms & Programming | (C3) Languages & Automata | (C4) Informatics Systems | (C5) Computer Science & Society |
|---|---|---|---|---|---|
| (P0) Interacting & Exploring | | | | | |
| (P1) Modelling & Implementing | Creating procedures | Formulating sequences for action | | Precise commands as input for control | |
| (P2) Reasoning & Evaluating | | | | | |
| (P3) Structuring & Interrelating | | | | | |
| (P4) Communicating & Cooperating | | Explaining instructions for action | | | |
| (P5) Representing & Interpreting | | Interpreting instructions for action | | Assigning parts of the robot | |

## Description

Programming via the unplugged approach – as is usual in this field – is initially introduced without the use of a real informatics system. However, a kind of "pseudo" informatics system is often used for unplugged teaching of programming. This is often a robot symbolised by children themselves or by a character. Basic programming constructs and simple commands on command cards can be visualised in different ways using symbols, text cards, puzzle pieces or similar.

Children can be given different tasks for the robot to complete, such as guiding classmates through a course (Dworschak, 2015) or controlling robots across a playing field. They have to solve these tasks by programming, which is done with the help of the command cards. The level of difficulty and the number of commands or programming constructs can be increased gradually. The sequences of actions formulated in this way are then checked by guiding the partner child through the course or the robot over the playing field using the commands.

*Figure 43.* Children programming a robot over the playing field (example from the project Computer Science at Primary Schools NRW, photo Kathrin Müller)

Another well-known example for understanding algorithms and specific commands is the "jam sandwich algorithm"[47]. Here, the task for the children is to instruct a robot (usually implemented by the educator or teacher) to spread jam on a slice of bread. Part of this task is to write an algorithm.

The big advantage of this unplugged approach is that it is inexpensive and does not require any technical equipment at the learning venue. This can also reduce the entry difficulties for educators or teachers, because they do not have to be afraid of technology failure, for example. Nevertheless, they need a basic knowledge of computer science in the field of programming in order to be able to carry out appropriate learning units.

In principle, we think it makes sense to follow the unplugged approach when starting with programming and to use an informatics system in a further step.

**Justification**

The essential concept to be discovered in this example is the concept of programming. A basic understanding of what programming is in the first place, when I am programming and how it basically works, is important to get along well in our digital world. Especially when it comes to adapting programmes to one's own needs, be it through specific setting parameters or macros.

With his quote: "Programme or be programmed", Rushkoff (2010) also underlines the significance of basic programming skills in our digital world. It can be assumed that basic programming knowledge contributes to a better understanding of digital systems and also increases skills in the field of end-user programming. These skills are becoming increasingly crucial in today's world if you don't want to

---

47 http://code-it.co.uk/unplugged/jamsandwich

have to call a technician for every setting change in the smart home etc. and rely on them to get it right.

As children are also increasingly encountering informatics systems that they can programme themselves, at least at the end-user level, the skills they acquire support them in using these systems safely.

*Example C:* **Solving tasks/puzzles algorithmically with an informatics system**

### Executive Summary

Children deal with given small tasks that are to be solved algorithmically with the help of an informatics system (e.g., app or robot) and whose solution can be implemented and tried out.

### Objectives

a) Application of the basic structures of programming as a sequence of elementary, unambiguous, mostly simple instructions (e.g., step forward, turn)
b) Introduction to the use of repetitions, conditional statements, procedures

### Heat map

| Content Domains / Process Domains | (C1) Information & Data | (C2) Algorithms & Programming | (C3) Languages & Automata | (C4) Informatics Systems | (C5) Computer Science & Society |
|---|---|---|---|---|---|
| (P0) Interacting & Exploring | | Trying out solutions in a targeted and guided way | | Precise commands as input for control | |
| (P1) Modelling & Implementing | | Modelling processes | | | |
| (P2) Reasoning & Evaluating | | | | | |
| (P3) Structuring & Interrelating | | | | | |
| (P4) Communicating & Cooperating | | | | | |
| (P5) Representing & Interpreting | | | | | |

### Description

Comparable to the unplugged example, children are to solve simple tasks by controlling a robot-like character with (usually movement) instructions in order to reach a certain goal. The instructions can usually be carried out immediately and are animated and followed with the help of the robot. These examples tie in with Papert's constructionism learning approach (cf. Section 2.3), with the aim of learning from specific objects.

**Specific design:**

- Create graphics based on the topic of Frozen:
  https://studio.code.org/s/frozen/stage/1/puzzle/1
- Create a game based on the topic of Angry Birds:
  https://studio.code.org/hoc/1
- "Paint" numbers using BeeBot
  http://barefootcas.org.uk/wp-content/uploads/2014/09/Bee-Bots-1-2-3-Activity-Barefoot-Computing2.pdf
- Solve tasks using Lightbot:
  https://lightbot.com/

**Justification**

See Example B (above).

### 5.1.3 Presentation and Transmission of Information

**Prioritisation: C1** (Information & Data) **and P5** (Representing and Interpreting)
Also addressed: P0 + P2 + P5/C1 + C4 + C5 and optionally: P3 + P4

This domain is illustrated with sample units that are likely to be more suitable for the upper age group of primary schools (3rd/4th grade). Experience from the InfoSphere school laboratory is available for these units and a corresponding sequence of learning units was tested at primary schools as part of the project "Computer Science at Primary Schools" in cooperation with the NRW Ministry of Education[48]. Similar competencies (possibly at a lower level) can be developed for younger children by using the same combination of Process and Content Domains and age-appropriate examples.

**Executive Summary**

Information is the third basic quantity of nature, next to matter and energy (see Wiener, 1948). Dealing with information leads to permanent social change and should already be known and understood by primary school children. The aim is

---

*48 https://www.schulministerium.nrw/informatik-grundschulen*

to give primary school children the opportunity to gain their first experience of the digital world.

In the process, children's everyday questions regarding data transmission are to be addressed and explained, but also abstract everyday problems, such as errors in data transmission and the possibility of recognising them, are to be worked out and made tangible for children. As the basis of digital data transmission, the binary representation of digital data in 0 and 1 forms the connecting element.

In the following exemplary implementation using four building blocks, the children's experiences are to be taken up as an introduction. Building on this, the topics of binary representation, data transmission and error detection can be dealt with.

**Objectives**

The aim of possible modules on the topic of information representation and interpretation is to give children an insight into the digital world. Children should realise that information can be represented in different ways. For use in the "digital world", all information must be binary coded, which is indeed happening increasingly: Texts, pictures, sound and video are represented by a set of "0" and "1" and are thus transferable and processable by informatics systems.

1st building block: Access to the topic
Using QR codes, children learn that information can be represented in different forms (letters, numbers, codes).

There are also other age-appropriate approaches, such as colouring in boxes in a distributed way, which, when put together, make a picture.

2nd building block (possibly in combination with mathematics): Number representation
Children get to know the binary code. They realise that all numbers from 0-15 can be represented with the four-digit binary code.

3rd building block: Data transmission
Children learn to transmit simple messages with the acquired knowledge on binary codes and also to decode received messages.

4th building block: Error detection
With the help of a control bit, children learn one way to recognise errors.

Learners

- describe their personal handling of informatics systems,
- explain the principle of data input and data transmission,
- justify the limitation to the states 'power on' (1) and 'power off' (0),
- explain how information can be converted into digital data,
- convert numbers, letters into binary code and vice versa themselves,
- explain the necessity of error detection mechanisms in data transmission and
- describe a given procedure for locating an error and apply it.

**Heat map (summary of the 4 units):**

| Content Domains / Process Domains | (C1) Information & Data | (C2) Algorithms & Programming | (C3) Languages & Automata | (C4) Informatics Systems | (C5) Computer Science & Society |
|---|---|---|---|---|---|
| (P0) Interacting & Exploring | Discover different forms of representation | Converting decimal numbers into binary numbers | | Informatics systems represent information in a way that is understandable to the user | Information is represented and must be (correctly) interpreted |
| (P1) Modelling & Implementing | | | | | |
| (P2) Reasoning & Evaluating | Difference between information & data | | | Informatics systems work by means of 'power on' (1) and 'power off' (0) | Some forms of representation are easier for humans to read, others are easier for informatics systems to read |
| (P3) Structuring & Interrelating | Representing information is possible in languages, images, … | | | | |
| (P4) Communicating & Cooperating | Data transmission | | | | |
| (P5) Representing & Interpreting | Representing decimal numbers and letters in binary code | Converting decimal numbers into binary numbers | | | Errors can occur during transmission or interpretation |

## Description

Possible specific ideas for implementation of a series of 4 units with the topics: Coding, Binary Representation, Data Transmission, Error Detection:

*Unit 1 – QR-Code*
In the first unit, children should recognise the relationship of the forms of representation discussed in the course to their own lives. QR codes serve as an introduction here. Children describe them (the central point is the pure colouring in black and white, i.e., two states) and discuss where they have already discovered them.

*Unit 2 – Binary Representation*
The aim of the second unit is for the children to realise that, in addition to the familiar decimal representation, there is also a special form of representation with which computer systems work: the binary representation. Even pre-school children are already familiar with different representations of numbers, e.g., fingers, lines, matches or apples. It is important that children recognise the comparison or analogies between the familiar ten-digit decimal system and the two-digit binary system. In both systems, only a certain number of different numbers can be represented with a certain number of digits (with three digits in the decimal system the numbers from 0 to 999 and correspondingly in the binary system only the numbers from 0 to 7). It is equally important that the educator or teacher establishes the reference to familiar informatics systems (computer, smartphone, tablet). The fact that computer systems work by 'switching power on and off' should be addressed and that data can be transmitted via a cable by (very fast) changes (see Unit 3).

To get a feel for this form of number representation, children could make a list of all possible combinations and thus realise that the digits 0 to 15 can be represented by the four-digit code.

It is also important that learners realise that the form of binary representation can also be transferred to other things with two states (lamp on/off, small card in blue/red).

*Unit 3 – Data transmission*
The aim of this unit is to extend the acquired knowledge about number representations to letters and punctuation marks and to experience ways for transmission.

The unit ties in directly with the previous unit by using a code table for the binary representation of letters and punctuation marks instead of decimal numbers. In this way, words or even sentences can be coded. It can either be written down by hand on paper and the message then passed on, or it can be done electrically using switches, cables and lights. Children can thus send messages to each other "like a computer". In doing so, the whole process starting from the information, to its binary representation, the transmission and finally the retransmission is repeated several times and consolidated. Here, too, the reference to real computer systems should be made.

*Unit 4 – Error Detection*
In this last unit, the aim is to make the learners aware of the problem of transmission errors and to give them a first idea (e.g., with the help of test bits) of how these errors can be detected.

A faulty message is sufficient to make the topic of 'transmission errors' intuitively accessible to the children (e.g., in the word KINDE/AR, an A (00001) could be transmitted instead of an E (00101)). The topic of transmission errors is addressed in a class discussion by means of such an example. The example of controlling a robot can also be used to illustrate that a transmission error can not only lead to a wrong letter in a message but can have far more serious consequences. Afterwards, the children can develop their own ideas for error detection. Finally, the unit leads to the possibility of a check bit, which can be made clear to the children by means of another digit in the code table, which is filled with 0 or 1 in such a way that the number of digits is even in each case.

**Justification**

The essential concept to be discovered is that **information** can be represented in different ways. For the digital world, it is decisive that it is sufficient to distinguish between two states and that any information can be encoded in this way. These two states can (simply) be transmitted in different ways. Binary representations can represent any text, numbers, colours, addresses, images, videos, structured **data** such as tables, sports scores, etc.

Through uniform coding, information can be easily **transmitted** and understood by different informatics systems "all over the world" and represented for people in their language and in a suitable format (e.g., pictures, graphics, tables) (coding, transmission, processing, decoding). The principle of displaying data in different representations, e.g., different natural languages, spoken or written word, pictures or symbolic representation, tables, football results, etc., is essential for communication and thus also important outside computer science.

**Everyday relevance** for children can be established by picking up familiar information representations, e.g., traffic light signals, QR codes, colour markings when assigning objects among siblings, etc.

### 5.1.4 Exploring and Structuring the Internet

**Prioritisation: C4** (Informatics Systems) **and P0** (Interacting & Exploring)
Also addressed: P2 + P3/C1 + C5

**Executive Summary**

We use the Internet every day. We look at websites, write emails, chat with our friends, etc. But how does it all work? How does the website get onto our computer?

With these questions, children could learn about the individual components and ideas behind the technology. One idea would be to divide the study of the topic into two large parts. In the first part, children learn in small groups about the different technologies and procedures that are important for the functioning of the Internet. Station learning is particularly suitable for this, whereby each station is dedicated to a specific topic and makes it tangible through small experiments. The stations should deal with the following topics:

- Converting decimal numbers into binary numbers (see Section 5.1.3)
- Transmission of data (see Section 5.1.3)
- The fast network in the background
- Breaking down and structure plan of a website
- How a DNS server works
- The client-server principle
- Structure of Internet addresses
- Security during data transmission

A quiz, for example, in which the learners repeat and thus consolidate their acquired knowledge, is suitable as a backup.

**Objectives**

The main goal here is that children recognise the rough structure of the Internet and develop a feeling for the fact that data is sent over the net and also stored outside their own computer (or tablet, smartphone). They recognise the necessity of logging on to particular websites. This basic knowledge about the technology of the Internet enables children in the future to rationally reflect on their behaviour on the Internet and to form their own opinion beyond scaremongering and glorification. In detail, the following objectives and topics can be addressed in the example:

- **Relation to children's own use of the Internet**
  Children should be made aware of their own current and future relationship to the Internet.

- **The Internet as an interplay of components**
  By breaking down the informatics system 'Internet' into its components, the learners can work out an overall connection.

- **Representation and transport of information in the form of data**
  The learning content on abstracting information (e.g., e-mails and web pages) in the form of data represents a fundamental principle.

- **Sequences of action and client-server applications**
  Client-server principle and the principle of a DNS server are to be taught as basic principles regarding the sequence of actions in the context of the Internet.

- **Security of data transmission and encryption**
  Already in primary school, children can be made aware of the problems and dangers of using the Internet.

- **Structure of an Internet address (URL)**
  The URL as a central element of the Internet should be taught with regard to its structure in order to make it possible to determine its validity.

### Heat map

| Content Domains / Process Domains | (C1) Information & Data | (C2) Algorithms & Programming | (C3) Languages & Automata | (C4) Informatics Systems | (C5) Computer Science & Society |
|---|---|---|---|---|---|
| (P0) Interacting & Exploring | | | | | |
| (P1) Modelling & Implementing | | | | | |
| (P2) Reasoning & Evaluating | | | | | Everyday significance of the Internet |
| (P3) Structuring & Interrelating | | | Structure of an URL | Structure of networks/the Internet | Data security |
| (P4) Communicating & Cooperating | Data transmission in networks | | | Internet addresses/URL | Sending mails |
| (P5) Representing & Interpreting | Binary representation | | | Client-server model | |

## Description

The following description contains ideas from the module "How does the Internet work?" of the InfoSphere learning lab.

To motivate the children, they could brainstorm on different possibilities of the Internet based on the question 'What can the Internet do?' In this way, they start by learning and reflecting on what skills and applications are possible on and with the Internet. Then they can move on to the question 'How can the Internet do all this?'

As a basis for working through the stations, the step from data to representation as binary numbers for transporting this data is shown. A light bulb that can be switched on and off with a switch can serve as an example from everyday life for switching electricity on and off. Building on this, the representation of a picture as a number chain is then explained as an example. Here, the image is reduced to the individual pixel, and this is then coded into a binary number. In this way, this unit could tie in with the previous example on the topic of binary representation.

In the further course of the unit, the children work in small groups at stations, working out one aspect of the Internet at each such station. Due to the content-related foundations laid in the first part, it is possible that each station of the tour can be used as a starting point for the work at the stations. In order to give the children an orientation for the work at the stations, they should be introduced beforehand and the respective starting points and processes should be briefly explained. In addition, the children could be given a worksheet on which they receive a stamp for each station worked on, so that they can keep track of the process themselves.

Throughout all stations, some models appear again and again. Thus, these model-like illustrations retain their meaning for the duration of the station learning. Binary numbers can be represented in the form of small wooden beads. Here, the colour white represents the number 0 and black the number 1. These beads are transported in different ways. Mainly, pipes are used as a model for the cable connections through which the beads can be sent.

Following the station learning, the newly acquired knowledge about the Internet could be consolidated in a quiz in the style of the German TV programme '1, 2 oder 3' (in the broadest sense similar to "Who wants to be a millionaire?"). In this quiz, questions are asked about the individual stations with three possible answers and the children have to stand in the field behind which they assume the correct answer. If the answers are correct, the children receive points. In this way, the competitive nature of this quiz is promoted and thus the motivation to perform is generated and maintained.

This part has two main purposes. The content learned in the stations is to be repeated and consolidated. The questions are therefore chosen to fit the learning objective of the respective station.

Further information on the InfoSphere module is available at http://schuelerlabor.informatik.rwth-aachen.de/module/internetspiel.

**Justification**

As a worldwide medium of communication and information, the Internet has comprehensive social significance. Skills and abilities and, most importantly, an understanding of these technologies have become indispensable for participation in social life. Above all, access to knowledge about society as a whole is increasingly taking place via the Internet. In the future, children will no longer be able to participate fully in social life (in everyday life as well as at work) without the appropriate prior education.

The smartphone is replacing the conventional mobile phone and the computer is more and more making its way into children's and young people's rooms. This is shown in a study by the Initiative D21 entitled "Education via the Internet: How connected are Germany's children?" (Initiative D21 e.V., 2008). In the age group of 7- to 10-year-olds, 86.2% of the children use the computer at home. Among 11- to 15-year-olds, 93.7% have a computer at home. Most of them (87.1%) use the Internet on the computer (see Initiative D21 e.V., 2008, p. 7).

### 5.1.5 Modelling of Automata

**Prioritisation: C3** (Languages and Automata) **and P1** (Modelling and Implementing)

Also addressed: P0 + P5/C4 + C5 and optional: P1/ C2

*Example*: **Exploring and modelling a traffic light system at a pedestrian crossing (intersection) as an informatics system**

**Executive summary**

The example describes the possibility of implementing competence domain P1/ C3. In the example, traffic education and aspects of computer science can be combined by exploring a real traffic light system. The example should help children to develop the ability to discover informatics systems in their real world and to recognise associated information-processing procedures. The example is suitable for introducing children to important concepts of computer science of finite automata and state-oriented modelling. At the same time, the social significance of such systems can be made tangible, e.g., by comparing it with traffic regulation by a human being.

## Objectives

The children should...

- be able to describe a traffic light system and its functions,
- be able to describe the significance of a traffic light system for traffic regulation,
- be able to describe the traffic light system as an informatics system with input and output devices,
- be able to describe the traffic light system as an automaton with input signs, output signs, states, rules for state transitions and a 'language' used for this purpose,
- be able to model a simple usage scenario (pedestrian traffic lights) as an automaton and
- (optionally) be able to programme a virtual traffic light simulation on the screen or by means of a model with a visual programming language (e.g., Scratch/Arduino).

## Heat map

| Content Domains / Process Domains | (C1) Information & Data | (C2) Algorithms & Programming | (C3) Languages & Automata | (C4) Informatics Systems | (C5) Computer Science & Society |
|---|---|---|---|---|---|
| (P0) Interacting & Exploring | Recognising red as information for stop, green for go | | Exploring the traffic light model for states, inputs and outputs | Investigating the traffic light system as an informatics system | Exploring a real traffic light system |
| (P1) Modelling & Implementing | Tram vs. car vs. pedestrian traffic lights | Optional extension: implementation with Scratch/Arduino | Modelling a traffic light system in a state-oriented way | | |
| (P2) Reasoning & Evaluating | | | | | |
| (P3) Structuring & Interrelating | | | | | |
| (P4) Communicating & Cooperating | | | | | |
| (P5) Representing & Interpreting | | | Presenting a graphical automaton model of the traffic light | | |

## Description

From the point of view of computer science, essential aspects of state-oriented modelling can be illustrated with this example. In contrast to the real world, an abstract model of an informatics system is created in the process of system modelling through context reduction (de-contextualisation), abstraction and formalisation.

The children can be engaged in the following activities (some of which are optional) in a pedagogical learning scenario:

- children explore how the traffic light system works at an intersection (possibly in combination with traffic education),
- the system can be reconstructed with paper or wooden blocks,
- application scenarios can be simulated (e.g., Playmobil characters or similar in role play),
- the status of the traffic light can be observed,
- interdependencies of the states of the traffic lights at the intersection can be observed,
- traffic lights can be regarded as finite automata with states and state transitions,
- traffic light intersections can be considered as automata with transitions,
- a pedestrian traffic light with input (pressed; not pressed) can be used as an introduction to state-based modelling,
- connection between 'language' with input and output alphabet can be established,
- other possible examples: railway barrier, construction site traffic lights,
- social aspect: comparison of traffic regulation by policeman and traffic light,
- optional extension for programming/simulation with Scratch on the screen or control of an intersection model with e.g., Arduino.

The starting point can be the exploration of a real pedestrian crossing controlled by traffic lights with a push button. In the further course of this didactic module, a picture (video) of the pedestrian crossing can also be used. The following further course of the learning scenario would be conceivable: From the photo of the pedestrian crossing, a drawing, a model made of cardboard with coloured tiles can be created. The scenario "crossing the road" can be acted out in a role-play or with characters. The states of the traffic lights and the push button can be entered in a table. State transitions depending on the push button, as well as input and output symbols (pressed, not pressed, red, yellow, green) can be acted out and written down. A variable timer (clock) can be taken into account as a pulse generator for state transitions. In this way, the principle of a finite automaton can be clarified as an abstract concept of computer science, which is clearly different from the technical realisation of an automaton (e.g., beverage dispenser).

Other simple examples would be, for example, a railway barrier, a construction site traffic light, an automatic door or a simple beverage dispenser.

Depending on the learning group, the example can be made more complex (intersection with traffic lights, intersection with pedestrian lights, intersection with contact threshold for cars). Optionally, automata can also be realised algorithmically, e.g., by creating a corresponding programme with Scratch and visualising it on the screen or by controlling a traffic light model with LEDs using an Arduino.

From the point of view of automata taking over human activities, traffic regulation by the police can be compared to that by a traffic light (or: beverage dispenser and salesperson).

This last example will also show how the criteria for the selection of Content and Process Domains described in Section 3.4 can be used to justify the prioritised selection of a Content Domain. Furthermore, the criteria can also provide important information on the subject didactic implementation of the example, taking into account the cognitive and motivational prerequisites of the learning group.

**Application of the criteria to the prioritisation recommendation P1/C3**

- **Subject-specific and subject-didactic concepts:** State-oriented modelling and the concept of finite automata are important concepts in computer science and are of great significance for modelling and understanding informatics systems.

- **Learning and developmental psychological aspects:** Children should be able to understand the process of abstraction and formalisation from the real informatics system to the abstract model (with assistance).

- **Relevance to daily life:** Traffic situations and traffic lights are important elements of children's everyday experience.

- **Motivation:** The functioning of a traffic light and the concept of an automaton can be explored in a motivating and playful way.

- **Subject-specific interest:** Exploring and designing basic functional principles of an informatics system and its inner states in a playful way can arouse interest in computer science.

- **Self-efficacy:** Modelling and, if necessary, (unplugged) programming of an informatics system familiar from everyday life, such as a traffic light system, can foster children's confidence in their handling of and ability to design informatics systems.

- **General education:** The children's confident handling of the informatics system 'traffic light unit', the acquisition of a conceptual understanding (automaton, internal view and external view of informatics systems) and initial experiences with automation processes primarily concern the following general education aspects: A2, A4, A6, A7, A9, A12 (see Section 3.4).

- **Overarching basic competencies:** Exploring the traffic light system, designing and explaining the model and testing application situations at the traffic light intersection in role plays can promote communication and teamwork skills.

- **Reference to didactic concepts:** Connections to traffic education and social studies and science in primary education (control of the traffic light model) can be made.

- **Existing practical experience:** Corresponding teaching models exist in basic computer science education; but with an older target group in mind.

## 5.2  Summary Heat Map of Priority Setting in the Examples

When the heat maps from the individual examples presented above are summarised, a recommendation emerges for prioritising competence fields that can be implemented in early information technology education in child-care centres and primary schools. However, this recommendation does not contain any suggestions for a sequence of the competence fields and thus for a sequencing of topics, since the fields addressed in the examples can be dealt with largely independently of each other. Thus, there is no hierarchical dependency of competencies to be acquired in these fields. However, the examples presented vary in complexity and when selected, early childhood educators and primary school teachers should take into account the cognitive abilities and motivational preferences of their learning groups.

## 5 Examples of Prioritised Competence Domains for Computer Science Education

| Content Domains / Process Domains | (C1) Information & Data | (C2) Algorithms & Programming | (C3) Languages & Automata | (C4) Informatics Systems | (C5) Computer Science & Society |
|---|---|---|---|---|---|
| (P0) Interacting & Exploring | | Exploring behavioural patterns, Conducting experiments (1: Interacting and exploring informatics systems) | | Orienting and exploring of components and functionality (1: Interacting and exploring informatics systems) | |
| (P1) Modelling & Implementing | Creating procedures (2: Algorithms and Programming) | Planning a sequence, modelling, formulating sequences of action (2: Algorithms and Programming) | State-oriented modelling of a traffic light system (5: Modelling of automata) | Precise commands as input for control (2: Algorithms and Programming) | |
| (P2) Reasoning & Evaluating | | | | | |
| (P3) Structuring & Interrelating | | | | Structure of networks/the Internet (4: Exploring and structuring the Internet) | |
| (P4) Communicating & Cooperating | Data transmission in the network (4: Exploring and structuring the Internet) | | | | |
| (P5) Representing & Interpreting | Representing decimal numbers and letters in binary code (3: Presentation and transmission of information) | Interpreting instructions for action (2: Algorithms and Programming) | | | |

*Figure 44.* Heat map summarising the priority setting of competence fields

# 6 Prerequisites for Successful Early Computer Science Education

## 6.1 General Conditions for Successful Implementation

Since only few research results are available, in the following, we refer to comparable conditions for successful early mathematical education and apply them to the field of computer science education, as presented by Benz et al. (2017).

Successful promotion of computer science education requires the fulfilment of various conditions, which, on the one hand, concern the competencies and attitudes of the early childhood educators and primary school teachers. On the other hand, they also include institutional and equipment-related framework conditions. The institutional framework conditions mainly concern the provision of time, space, organisational and equipment-related conditions.

In summary, especially as a conclusion from the explanations in Chapter 4, it can be stated that early childhood educators and primary school teachers:

a) should have professional competence, in particular to understand the teaching-learning material offered and to plan and design their own learning units on this basis,

b) need pedagogical-didactic competence to recognise and use learning offers of computer science education and to select, adapt and, if necessary, further develop the offered materials in a learner group-oriented manner,

c) have positive attitudes towards computer science but should also be open to new perspectives on existing ideas.

### 6.1.1 Educators' and Primary School Teachers' Subjective Theories and Attitudes Towards Computer Science

So far, there are only few findings on the influence of attitudes towards computer science and of subjective theories on the acquisition of computer science competencies. However, based on the findings from other subjects, it can be assumed that these have a direct influence on how and with what enthusiasm teachers and educators teach a subject.

Last but not least, the topic is also influenced by mental barriers and fears stemming from one's own experiences in dealing with digital media. Knobelsdorf and Schulte (2007) identified attitudes that emerged among first-year students as "insiders" or "outsiders", which led to "outsiders" limiting themselves primarily to the use of computer programmes and essentially accepting problems, while "insiders" saw the computer and thus computer science as a tool that could be shaped.

Since it can be assumed that many educators and teachers in child-care centres and primary schools have incomplete and incorrect ideas about computer science and accordingly about computer science education, they should be open to a change of perspective and their own learning experiences. Ideally, they should be only one 'step away' from changing from "outsiders" to "insiders".

Since creativity is increasingly becoming a goal of computer science education, the aim should be to design computer science lessons in line with the creative potential (see Romeike, 2008).

## 6.1.2 Subject-Didactic Competencies of Early Childhood Educators and Primary School Teachers

Everyday life offers many opportunities for computer science education, not least because children's lives are increasingly shaped by digital artefacts (cf. Section 2.1). Accordingly, Borowski, Diethelm and Wilken asked third and fourth graders what questions they would like to have answered on the topic of "computers, mobile phones, robots, etc." (Borowski, Diethelm & Wilken, 2016). The resulting questions cover a wide range of computer-related topics: From the history, future and function of informatics systems to security and legal aspects. Such intrinsically anchored questions offer potential for learning opportunities in computer science education. However, it is a challenge for educators and primary school teachers to recognise the potential behind the children's questions and to identify suitable everyday situations for designing learning situations. There is also a danger that this view links learning opportunities in computer science exclusively to informatics systems and, furthermore, creates a distorted picture that fundamental ideas and concepts that have no direct relation to informatics systems are neglected.

Analogous to the requirements formulated in the expert report on mathematics, early childhood educators and primary school teachers have demanding tasks (Benz et al., 2017):

1. From the existing *teaching-learning materials* for computer science education, but also from other contexts (cf. Section 2.3), materials must be selected that have a high computer science potential and correspond to the current developmental level of the children.
2. *Learning situations* must be created or addressed so that processes of computer science education can be implemented with the available material.
3. Digital artefacts, but also unplugged materials for computer science education have an inherent *challenging character* that should arouse children's curiosity and interest in engaging with the subject matter. Educators and teachers should support this, e.g., by embedding these artefacts and unplugged materials in a play situation.
4. The *children's motivation* should be addressed by the educators and teachers, showing their own interest in the subject matter and inspiring the children with their enthusiasm for the respective subject.
5. By *creating communicative situations*, the children's reasoned exchange about contexts, assumptions and conclusions should be practised, and reflection on what they have experienced should be strengthened. For this, it is necessary that the technical terms of computer science are used confidently but also in an age-appropriate form.
6. Based on experiences with computer science education processes of older children and adolescents, it can be assumed that gender-specific differences in interest and handling of technology and computer science develop with increasing age, which do not yet exist or are less pronounced in child-care centres and primary schools. In this respect, it is the task of the educators and teachers not to reinforce possible prejudices, to reflect on their own attitudes and stereotypes and to motivate and consider boys and girls equally in the teaching-learning situations.

### 6.1.3 Cooperation Between the Educational Institution, Family and Decision-Makers

Cooperation between educational institutions and families is also essential in the field of computer science competencies. The aim should be to agree on common values and goals for computer science education, even if this is more likely to be the result than the prerequisite of the educational process. It can be assumed that many parents have unclear ideas about what the goals and potentials of computer science education in child-care centres and primary schools actually are. Some legal guardians may also have no knowledge about computer science or have a different perception of computer science that manifests itself in other ways; all in all, there will probably be very different notions of computer science. Thus, it would be detrimental if situations for computer science education are demotivated in advance from home, e.g., that "the child should not play computer games" or that false expectations are raised ("at child-care centres, children are taught how to use computers").

This also goes hand in hand with the fact that children have very different ways of dealing with digital media at home, ranging from rejection to extensive use (cf. Section 2.1). It would therefore be desirable for the cooperation between the educational institution and the parental home to also include information and cooperation about how digital media can be used not only receptively but also creatively and how computer science education can contribute to this. For early childhood educators and primary school teachers, this results in an important field of action to make computer science education in child-care centres and primary schools transparent for parents and to involve them productively in the computer science learning processes in the educational institutions. A transparent design of concepts of computer science education by early childhood educators and primary school teachers is also important in order to convince the decision-makers involved in the educational institutions of these concepts and to win them over for a design of the organisational framework conditions conducive to learning.

### 6.1.4 Organisational Conditions in the Educational Institution

There are different ways of organising learning opportunities for computer science education: Some previous experience has been gained with group work, partner work, in a formal organisational setting or as a working group, as a performance or participatory show (cf. unplugged shows) or also in the context of an exhibition. In the child-care sector, learning processes for computer science education should be designed according to the child-friendly, playful work formats that have been tried and tested there.

For primary schools, it is recommended to organise computer-science-related teaching projects. In order to sustainably promote computer science teaching, efforts should be made to create a learning area, a separate subject or to integrate computer science into social studies and science in primary education and to give subject teachers the opportunity to qualify. It should be taken into account that the approach of integrating content of computer science into other subjects has failed for various reasons, at least in the domain of junior secondary education (see e.g., Breier, 2004b).

### 6.1.5 Equipment-Related Requirements

Computer science education can in principle also take place without informatics systems: For example, aspects of the goal domains 'algorithms', 'languages and automata' or 'information and data' can be taught unplugged. Also, many informatics systems can be found in everyday life today, so that they can be observed and analysed together. Nevertheless, the use of informatics systems offers opportunities for motivation and learning experiences that would hardly be possible without them. For example, running an algorithm on an informatics system can illustrate the need for uniqueness and executability of algorithms in a comprehensible and credible way. Creating their own game, which they can proudly show to their friends or parents, gives children an idea of the creative possibilities in dealing with computer science. Section 2.3 outlined various possibilities that show specific ways of implementing computer science education with informatics systems. Of course, this also presupposes that the educational institutions have the appropriate equipment, but this need not be seen as an indispensable prerequisite for successful informatics education, especially in child-care centres.

## 6.2 Measuring Instruments to Determine the Prerequisites for Success

For the field of computer science education in elementary and primary education, there are no empirically-based models of competence available so far. This results in a lack of reliable instruments for measuring the competencies acquired by children in the elementary and primary sector of computer science education and, associated with this, an empirically sound evaluation of concepts of computer science education in child-care centres and primary schools, which is an essential element for their successful implementation in the pedagogical practice of child-care centres and primary schools.

Not least because the scientific monitoring of the foundation's work should be multi-layered and further research on early computer science competencies should be promoted. Based on the competence models partially developed for computer science in senior secondary education and teacher training (cf. Chapter 4), various fields of activity can be identified for the evaluation of learning processes and conditions for the success of computer science education in child-care centres and primary schools, for which suitable domain-specific measurement instruments would then have to be developed.

### 6.2.1 Conceptional Evaluation

The proposed measures and examples should be subjected to critical analysis by further experts in computer science didactics as well as by experts in early childhood and primary school education.

### 6.2.2 Material Evaluation (Analogous to Mathematics Didactics)

The materials developed by the "Haus der kleinen Forscher" Foundation for early computer science education should be subjected to a critical analysis by a group of people from computer science didactics, relevant neighbouring disciplines and representatives from the field. Regular specialist forums on computer science with appropriate experts from didactics and those with practical experience, who are familiar with the work of the Foundation, could be helpful. In addition, the quality of the material should be continuously improved through the interplay of theoretical consolidation, development work and empirical, mostly qualitative research in the paradigm of design research.

Furthermore, research projects on the impact and application of the materials should be established with the following target groups:

- children
- early childhood educators and primary school teachers
- multipliers

### 6.2.3 Evaluation of Measures

Individual concepts of computer science education in child-care centres and primary schools should be evaluated, with the following target groups in mind:

- children
- early childhood educators and primary school teachers
- multipliers
- parents

### 6.2.4 Research on Effectiveness

Research on effectiveness is linked to the evaluation of the measures and deals with the following subjects of investigation, with special attention to the following target groups:

- Children
    - A: Non-cognitive competencies (motivation, interest, self-efficacy, attitudes) → Here, for example, there are already studies on gender differences (cf. Chapter 2).
    - B: Specific subject-related competencies related to a teaching module (module-related analysis of learning outcomes).

- C: General computer science competencies → e.g., recognising computer science phenomena, recognising informatics systems and how they work etc.
- D: Evaluating learning processes (e.g., in terms of overcoming learning barriers/transition of learning concepts, e.g. from unplugged to plugged).

■ Early childhood educators and primary school teachers:
- A: Attitudes/curricular beliefs

    The importance of teachers' attitudes, stances and understanding of their professional role for their pedagogical practice has long been demonstrated in various empirical studies (cf. Chapter 4; e.g., Eulenberger, 2015). In her study on the professional identity of 'CS teachers', Ni (2011) has specified the influence of 'beliefs' on the teaching practice of computer science teachers. Among other things, she identified the two general influencing factors, 'perception of CS' and 'perception of teaching'. In addition, the attitude or general acceptance of technology in the classroom plays a role (as applies to children, see above); e.g., Gil et al., 2007). Here, too, existing instruments can be adapted (e.g., TAM 2.0 (Bagozzi, 2007) or INCOBI-R (Richter & Naumann, 2010; Richter, Naumann & Groeben, 2001).

    Possible survey criteria in this area are:
    - Attitudes towards informatics systems
    - Use of informatics systems in everyday life and at school
    - Attitudes towards computer science in primary schools
    - Curricular beliefs/curricular emphasis

- B: Subject-didactic competencies

    In order to evaluate the subject-didactic competencies or the acquisition of competencies by early childhood educators and primary school teachers, competence measurements should also be conducted in this target group. The test instruments items could be adapted for the target group from theoretically and empirically-based competence models for the education of computer science teachers and derived from measurement instruments already available there (cf. e.g., KUI, CoRe in Chapter 4: Bender et al., 2015, 2016; Buchholz et al., 2013; Linck et al., 2013; Magenheim et al., 2010; Neugebauer et al., 2015; Williams & Lockley, 2012) and refer to subject-specific, subject-didactic, motivational and volitional aspects of competence (Weinert, 2001). This also includes the assessment of practice-related action competence in computer science-related learning situations through observations.

- Parents:

  As already described in Section 6.1.3, the parents and the family environment of the children presumably have a significant influence on children's attitudes towards informatics systems. Therefore, it makes sense to survey indicators of the children's computer science-related family environment and the parents' attitudes towards the field and informatics systems as relevant factors of the conditions for the success of its education in child-care centres and primary schools. Specifically, this includes the following aspects:
  - Possible uses:
    Information and communication systems in the home environment (existence, forms and frequency of use)
  - Attitudes towards:
    - informatics systems
    - the use of informatics systems in everyday life and at school
    - computer science in primary schools
  
  As with the survey of teachers' attitudes, similar, partially adapted instruments should be used to investigate whether parents' attitudes and use of IT systems in the home have an influence on the acquisition of computer science competencies and children's attitudes towards computer science and informatics systems.

# 7 Conclusion

With this report, the team of authors has broken new ground both from the perspective of subject didactic research and from the perspective of the pedagogical practice of computer science education in the child-care and primary school sector. There are several reasons for this, including the fact that there is still little subject didactic research on computer science education compared to other established subjects, as well as the comparatively weak anchoring of computer science education as a subject in secondary and primary education, especially in German schools. Against this background, the team of authors has attempted to work out important goals and conditions for successful computer science education in child-care centres and in the primary school sector. To this end, the concept of computer science education, its relationship to the technical discipline of computer science and the similarities and differences to digital (media) literacy were first characterised. In addition, existing, especially international approaches to computer science education in child-care centres and primary schools, as well as suitable learning environments and software tools were analysed for their usability in early computer science education. Relevant curricula were evaluated with regard to their references to standards of computer science education for the purpose of comparability, and in this way a category system that was at least partially empirically based, was developed to describe fields of competence in computer science education. From this analysis of existing international curricula for early computer science education and the basis of relevant concepts of different didactic approaches in computer science, a new Process Domain, 'Interacting and Exploring' with and of informatics systems was introduced in comparison to the GI educational standards. Its significance for the playful, explorative handling of

informatics systems in child-care centres and primary schools was discussed in detail. In this context, reference was also made to relevant studies on the necessary cognitive prerequisites and development of children, which are considered relevant for the acquisition of basic competencies of computer science education. Attitudes, motivation and beliefs of early childhood educators and primary school teachers towards computer science and computer science education were described as important prerequisites for the acquisition of their subject-didactic competencies in the field of early computer science education. All in all, well-trained early childhood educators and primary school teachers were characterised as an essential factor for the successful implementation of concepts of computer science education in child-care centres and primary schools. Finally, important criteria for the construction of empirical measurement instruments for the evaluation of conditions for success and of pedagogical practice of early computer science educations were derived as a conclusion of these explanations. The authors themselves consider the practical testing of the examples and the concepts presented, as well as an accompanying formative evaluation, to be an essential factor for the successful implementation of early computer science education in the practice of child-care centres and primary schools.

With this report, the team of authors hopes to support the important activities of the foundation "Haus der kleinen Forscher" in the field of early computer science education and thus to make a modest contribution to the implementation of computer science education in child-care centres and primary schools.

# B  Professional Recommendations for Informatics Systems

Nadine Bergner, Kathrin Müller

1  Introduction
2  Overview of Possible Informatics Systems
3  Description and Technical Assessment of Individual Informatics Systems
4  Recommendations
5  Conclusion

# 1 Introduction

Informatics systems are playing an increasingly important role in our digital world. These are no longer exclusively large informatics systems operated by experts but countless small and larger everyday devices such as smartphones, washing machines or cars. In some cases, these systems are clearly recognisable as informatics systems (as in the case of a laptop or smartphone), while in other instances (like, for example, with coffee machines), it is not always immediately obvious that an informatics system is hidden inside. Even some (toy) devices for toddlers (e.g., baby monitors and talking dolls) contain a complete informatics system. These examples show that children come into contact with informatics systems long before they ever sit in front of a computer. Probably the most widespread and already very complex informatics system is the smartphone. Here it is noticeable that children already handle the devices completely intuitively at toddler age, although they were primarily developed for adults. That is where the generation derived its name from: digital natives.

However, if the intuitive operation is not purposeful, it quickly becomes clear why sound computer education is crucial for this generation in particular. Children often copy processes that they have observed with their parents; but they do not recognise the underlying logic. They do not understand, for example, whether files are stored locally on a device or distributed on the Internet, what is public and private, how the networking of devices works and how this affects our lives.

Therefore, children should learn what it means to work with informatics systems already in kindergarten and primary school. They should not only use digital devices but also understand that these devices do what they are "told" to do. Therefore, it is important to teach them not only the "how" but also the "why". For the "why", a basic understanding of informatics systems and how they work is essential. The main goal for children in kindergarten and primary school age is to gain first experiences in the field of computer science in order to develop a basic understanding of computer (systems) in the long run. Only in this way can children become responsible co-creators (and not just passive users) of our digital environment.

## 2 Overview of Possible Informatics Systems

Today, there are numerous informatics systems for use in elementary and primary education. These systems can be categorised in different ways. On the one hand, there are systems that can be used autonomously, such as Cubetto, Beebot or Kibo (more information on these systems follows in the next chapter). Other systems can additionally be used in conjunction with an app, such as the Ozobot, or a computer, such as Lego WeDo. Systems such as language Scratch or Scratch Junior do not require any additional hardware other than a tablet or computer.

But which system is suitable for which age? Which previous knowledge is required of the children – and also from their learning supporters? And which computer science goals can best be pursued with which system?

This expert recommendation addresses these questions and attempts to show the current state of existing informatics systems for children[49]. The assessments of user-friendliness, previous experience, etc. are based on our experience as well as on manufacturer information.

The informatics systems described below are ordered by level of difficulty or complexity.

*Systems with low complexity:* These systems can usually be used independently and often require low IT competence of the learning supporter.

- Cubetto
- Beebot
- Kibo
- Ozobot without visual language

*Systems with medium complexity:* These systems are often used in conjunction with an app (for smartphone or tablet) or also with a computer. Since a programming language is used, the learning supporters require computer science skills in programming, depending on the system.

- ScratchJr
- Dash & Dot
- WeDo
- Makey Makey

---

[49] This article is a translation of the German version, dated 2018.

- Ozobot with visual language (the special feature here is that this system can cover the entire range of complexity)

*Systems with higher complexity:* these systems require a higher level of expertise from the learning supporter:

- Scratch
- Lego Mindstorms
- Arduino-Micro-controller with ArduBlock

# 3 Description and Technical Assessment of Individual Informatics Systems

## 3.1 Cubetto Robot from Primo Toys

**Overview**

The Cubetto robot by the company Primo Toys was developed in 2013 with the aim of introducing children aged 3 years and older to the world of programming in a playful way. Cubetto is a cube-shaped wooden robot with a smiley face and two wheels to move around. In fact, moving is all Cubetto can do. The special feature of this system is that it comes with its own programming board, which is also made of wood. Children can insert the four differently coloured and shaped wooden puzzle pieces (corresponding to the individual programming commands/ movement instructions) into this board. An arrow on the board indicates the order for these actions. Once the children have completed their programme, consisting of up to 12 commands (puzzle pieces), this is transferred to the robot by pressing the single button. The robot processes the entered commands one after the other. At the same time, small lights on the programming field indicate which command is currently being processed.

The robot and programming board come with a playmat on which places are marked in the form of simple graphics (e.g., a castle tower or sailing ship) so that children can create tasks for the robot in the style of "Go from the castle to the lake". However, learners can also develop and solve tasks in a room, such as e.g., "drive around the chair".

The learners can extend this by programming a so-called method. This is an IT construct to streamline programme codes by not programming repetitive elements more than once. For this purpose, up to four commands are outsourced to a separate area (outline). In addition to the "forward", "left" and "right" blocks, there is another block that invokes the method and thus triggers the execution of the up to four commands. This additional building block can

now be inserted at several points in the main programme, where the same method is then executed again and again. All in all, the limited range of commands and the maximum programme length result in a quite manageable application area, which distinguishes Cubetto as a beginner's tool.

**Target Group**

According to our experience so far and the application scenarios presented on the distributor's website, Cubetto is particularly suitable for children from pre-primary age (especially with Montessori pedagogy), as it corresponds to a wooden toy and children can explore it without instructions. Most children are already familiar with the goal of putting wooden blocks into the matching holes from pre-primary age. Learners are also familiar with pressing a button. The only requirement for the learning supporter is to trigger a thought process in the children about the connection between the programme and the execution of the robot, as the transmission is wireless and therefore not immediately obvious to the children.

**User-friendliness**

This robot system is extremely user-friendly and can be applied regardless of the child's or learning supporter's previous knowledge. All the learning supporter needs to do is insert batteries into the robot and the programming board in advance and switch both on using a small slide switch. The connection is established automatically, so that programming can start immediately. No other IT system (computer, laptop, tablet) is required.

**Previous experience and knowledge required of the learning supporter**

As described above, this system can be used without any previous computer science or technical knowledge. Only the idea of the linear sequence of commands in the main programme and the method construct need to be understood in advance. Apart from this, Cubetto only requires creative tasks, which, however, the children develop themselves with great enthusiasm after an initial exploration.

**Previous experience and knowledge required of the child**

Again, no previous knowledge is absolutely necessary. The idea of putting wooden puzzle pieces into matching holes should be understood. The topics of robots, programmes, commands etc. can be covered with this learning tool.

**Learning objectives**

The overarching learning goal is to create a basic understanding of automatically running programmes. The children learn that programming means that a human enters a sequence of commands into an informatics system, in this case the robot, which then executes the programme at the push of a button. In detail, the children thus learn about algorithms and how to develop them themselves. Moreover, since Cubetto is ideally suited for being programmed by several children together, they also learn indirectly about goal-oriented communication via programmes and the challenges of teamwork.

**Reference to the goal dimensions of IT literacy**

The Cubetto robot set primarily addresses the "C2 Algorithms & Programming" *Content Domain*. Its target group-oriented design enables children of pre-primary age to develop algorithms themselves. Depending on the design of the discussions in the learning groups or further learning phases, features under "C4 Informatics Systems" and "C5 Computer Science and Society" can also be addressed.

In the *Process Domains*, the focus is on "P0 Interact and Explore", in that the programmable robot can be explored very freely by children. In addition, the learners can already implement their own programmes (an aspect under "P1 Modelling & Implementing"). When playing in teams, Process Domain "P4 Communicating & Cooperating" is also applied.

## 3.2 Beebot from Terrapin

**Overview**

Beebot[50] is a programmable floor robot in the shape of a bee, for children in preschool and primary-school age. Its simple and child-friendly design makes Beebots a good first introduction to programming. You can programme the Beebot's moving sequence. For this purpose, there are seven buttons on the "bee". Beebots can move forward or backward (by exactly 15 cm each), and perform a 90-degree turn to the left or right. By pressing the direction buttons, up to 40 consecutive commands can be programmed as a sequence. The "Go" button in the middle starts and executes the entered sequence. In addition, there is a button to delete the previous sequence. If a sequence is not deleted, the following commands are added to the previous sequence. This enables children to work their way forward step by step. The Beebot's sounds or light signals help the children to understand their entries. For example, it flashes and makes a sound when a command of the entered programme has been executed.

The Beebot is powered by a rechargeable battery that can be charged via a USB interface.

There is a wide range of instruction materials for the Beebot, available from the manufacturer as well as from various research and educational institutions.

---

*50  https://www.b-bot.de/produkte/bee-bots/*

Many of these materials emphasise interdisciplinary learning and problem solving.

In addition to the Beebot, there are now other variants such as the Blue-Bot or the Pro-Bot. The Blue-Bot can additionally be controlled via a tablet. The Pro-Bot has a brand-new racing car design and offers innovative functions and thus more possibilities than the Beebot.

**Target Group**

The Beebot is suitable for children in pre-school and primary school, aged 4 to 9 years. For older children, the Beebot quickly appears too childish due to its bee design.

**User-friendliness**

The Beebot is a user-friendly system. The meaning of the control buttons quickly becomes clear, even without instructions, at least for older children; younger children may need a brief explanation.

**Previous experience and knowledge required of the learning supporter**

The only prior knowledge the learning supporter needs to have, is how the Beebot works and an idea of the linear sequence of commands. Once these aspects are understood, only suitable (creative) tasks are needed that are to be solved with the help of the Beebot. These can be simple path finding but also putting words together on letter mats. After the exploration phase, however, the children will certainly find their own first challenges. In addition, a wide range of application materials for the Beebot is already available, for example, on the manufacturer's homepage (https://www.bee-bot.us/bee-bot/beebot-curriculum.html).

**Previous experience and knowledge required of the child**

The children do not need any previous knowledge. It may be necessary to explain how the control buttons work. However, this depends on the age and previous experience of the child. For programming the Beebot (depending on the complexity of the route to be programmed), the children need a certain level of working memory. The route must be planned in advance, as the inputs in the Beebot are not visible and therefore cannot be followed directly. However, to support this, there are maps with which routes can be pre-planned and then entered.

**Learning objectives**

The overarching learning goal is to create a basic understanding of automatically running programmes. As with Cubetto, the children learn that humans control an informatics system by entering commands (programming) and that the system then executes the programme. This way, the children learn how to use a simple informatics system and also how to solve problems with the help of algorithms. Since the Beebot is ideally suited for being operated by several children together, they also learn indirectly about goal-oriented communication via programmes and the challenges of teamwork.

**Reference to the goal dimensions of IT literacy**

The Beebot also primarily addresses *Content Domain* "C2 Algorithms & Programming". Due to the target group-oriented design, even children of pre-primary age can use the Beebot to develop algorithms in the context of problem solving and enter them into the Beebot (*Process Domain* "P1 Modelling & Implementing"). Depending on the design of the discussions in the learning groups or further learning phases, features under "C4 Informatics Systems" and "C5 Computer Science and Society" can also be addressed.

In the other *Process Domains*, the focus here is on "P0 Interacting & Exploring", in that the Beebot can be explored very freely by children and they can apply or develop their interaction skills when dealing with the system. When playing in teams, Process Domain "P4 Communicating & Cooperating" is also applied.

## 3.3 KIBO from KinderLab Robotics

**Overview**

The KIBO[51] is an educational robot kit designed to introduce children to the concept of programming in an interactive way. The KIBO robot is programmed using wooden blocks that are labelled and can be plugged into each other.

KIBO offers children numerous creative possibilities. The existing building elements offer many different design options, e.g., as a carousel, a robot with an ear, an eye, lamps or free design with additional handicraft material, in line with the children's imagination. This enables a wide range of possible uses, from specific tasks set by the learning supporter to free creative work with the KIBO.

---

51  *http://kinderlabrobotics.com/kibo/*

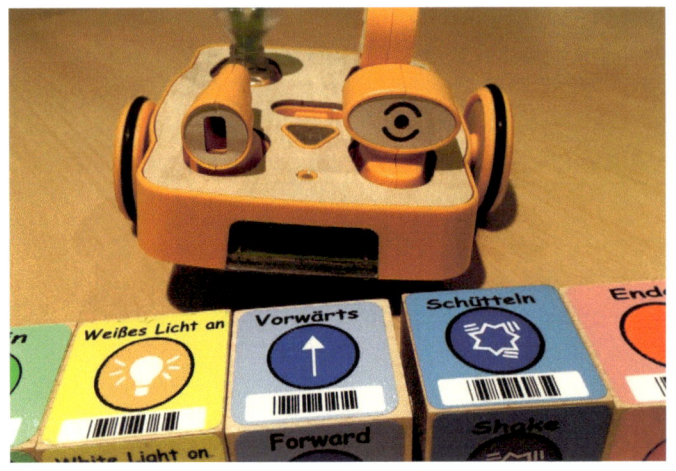

The KIBO is equipped with sensors, lights, motors. For example, it can react to clapping, its lights can shine differently and it can move in different directions.

As already mentioned, the KIBO is controlled with wooden blocks. On these blocks are instructions for the KIBO, which are sequentially lined up with a plug-in connection. They are then read with an infra-red scanner, which is located on the KIBO itself. Finally, the programme is started at the push of the button and executed by the KIBO. The instructions refer, for example, to the KIBO's direction of travel, the colour of the lamp, reactions to claps, etc.

Even though the blocks are lined up sequentially, there are already existing constructs such as a repetition and conditional instructions.

The instructions on the wooden blocks are presented both graphically and in writing, so that even children without reading skills can make their first programming experiences with KIBO.

**Target Group**

According to the manufacturer, KIBO is intended for children aged 4-7. From our experience, we would recommend using it at pre-school or primary school age.

**User-friendliness**

KIBO is very user-friendly and can be applied regardless of previous knowledge of the children or the early childhood educators and primary school teachers. The learning supporter only has to insert the batteries into KIBO in advance. The individual elements of KIBO can be plugged into KIBO quite intuitively. For smaller children, this step could also be carried out by the learning supporter beforehand. The only thing required for programming is plugging the wooden blocks together. They are then read with the infra-red scanner, which is located on the KIBO itself. Children quickly recognise how to handle KIBO correctly. In case of errors, it immediately gives feedback. After successful scanning, it starts at the push of a button.

**Previous experience and knowledge required of the learning supporter**

Basic prior IT knowledge is required to use the KIBO. Knowledge of linear sequence of commands, as well as functions from simple loops to conditions should be available to support the children if necessary.

**Previous experience and knowledge required of the child**

Children do not necessarily need any previous knowledge. They should understand the idea of putting wooden blocks together appropriately. The topics of robots, programmes, commands etc. can be covered with this learning tool.

**Learning objectives**

The overarching learning goal with KIBO is also to create a basic understanding of automatically executable programmes. Children learn the first basic concepts of programming and, above all, that programming means entering a sequence of commands into an informatics system, which the system then executes. In addition to the sequential order of commands, children learn about conditions and loops in a playful way. Moreover, since KIBO is ideally suited for being programmed by several children together, they also learn indirectly about goal-oriented communication via programmes and the challenges of teamwork.

**Reference to the goal dimensions of IT literacy**

KIBO primarily addresses the "C2 Algorithms & Programming" *Content Domain* as well as the "P1 Modelling & Implementing" *Process Domain*. Due to the target group-oriented design, KIBO can be used to develop initial algorithms and implement them with the help of the wooden blocks. Depending on the design of the discussions in the learning groups or further learning phases, features under "C4 Informatics Systems" and "C5 Computer Science and Society" can also be addressed.

In the *Process Domains*, another focus is on "P0 Interacting & Exploring", in that children can explore the programmable robot and its elements very freely. When playing in teams, Process Domain "P4 Communicating & Cooperating" is also applied.

## 3.4 Ozobot/Ozobit from Evollve Inc.

**Overview**

The Ozobot[52], introduced in 2014, is an approximately 2.5 cm tall robot. An Ozobot can both follow lines and – in the new generation, the Ozobot Ozobit[53] – move freely around the room. In addition, it can be programmed in two different ways, making it suitable for use at different age levels. The Ozobot is powered by a rechargeable battery that can be charged via a USB interface. There is a five-colour sensor on the underside of the robot that enables it to detect different colours on the floor. Based on lines and colours on the ground, the Ozobot can move in different directions at different speeds and even glow in different colours.

The small robot can be programmed in two different ways:

*Programming with colour codes*
It can be programmed with colour codes without any additional IT system. Besides the Ozobot, only pencils and paper are needed. The line sequence and colour codes are simply drawn on paper. With the help of these codes, the Ozobot can change its colour, speed or direction. So-called "SpecialMoves" also allow the Ozobot to turn or zigzag. In this way, entire courses can be built for the Ozobot. There are no limits to creativity. Users can also compile lines and colour codes on a tablet and let the Ozobot drive on them.

*Programming with the visual programming language OzoBlockly*
The new generation of Ozobots, called Ozobit, can be programmed with the visual programming language OzoBlockly[54]. It is similar to other visual programming languages such as Scratch (see below) but was developed specifically for the Ozobot. It can be used via an Internet browser on the computer, via a tablet or even a smartphone. The created programme can be transferred to the Ozobot by simply holding it up to the screen or placing it on the tablet.

The structuring into different levels is very useful. At the beginning, only programming blocks with symbols are available, so that even children without reading skills can programme the Ozobot. In the further levels, more and more complex programming blocks are added. Both the Ozobot's behaviour on lines and colours and its completely free movement across the room can be programmed.

---

52 http://ozobot.com
53 Since the Ozobit offers more possibilities than the first generation Ozobot, we will hereinafter always refer to this new generation Ozobit.
54 https://ozobot.com/create/ozoblockly

The motors of the Ozobot are very precise, so that tasks with measurements or synchronous applications etc. are also possible.

Furthermore, the Ozobot itself offers creative design possibilities via design options of the protective covers with paper or Ozobot-specific accessories.

All in all, Ozobot is an inspiring little robot with many creative possibilities and different levels of difficulty for young and old.

**Target Group**

Ozobot is suitable for use in primary school and beyond. The grade in which it is introduced depends on the design of the programme. According to the manufacturer, Ozobot is suitable for children from the age of 8. Because of the different possibilities for programming and thus the different levels of difficulty, there are no upper limits for the use of the Ozobot.

The suitability of the Ozobot for younger children would have to be tested. The meaning of many colour codes can easily be represented by symbols instead of writing, so that the Ozobot can also be used by children with no or very poor reading skills.

**User-friendliness**

As such, the Ozobot is very user-friendly. However, when using it, care must be taken that it is calibrated regularly. If it is used with non-original pencils, it must be ensured that the colour tones are very close to the original, otherwise the colours will not be recognised 100% correctly.

There is a good, detailed manual and many examples for the Ozobot. Furthermore, the corresponding JavaScript code can be displayed.

The disadvantage of OzoBlockly, including the manual, is that it is currently only available in English.

**Previous experience and knowledge required of the learning supporter**

When it comes to the necessary previous experience or knowledge of the learning supporter, a distinction must be made between how the Ozobot is to be used.

When using the Ozobot with paper and pens, no previous knowledge of the learning supporters is required. They must charge the Ozobot in advance via the

USB and should, of course, have already tried out the Ozobot themselves. The manufacturer of the Ozobot already provides material on the meaning of the colour codes. In addition, there are some worksheets and many creative ideas for using the Ozobot on the Internet.

Depending on the level of difficulty, previous knowledge of basic programming structures should be available for the use of the visual programming language in order to be able to support the children appropriately in case of difficulties or questions.

**Previous experience and knowledge required of the child**

Children do not need any prior knowledge to control the Ozobot via colour codes.

If the Ozobot is to be programmed with OzoBlockly, children should be confident in using the computer or tablet. However, previous experience of programming is not necessary.

The topics of robots, programmes, commands etc. can be covered with this learning tool.

**Learning objectives**

The learning objectives for using the Ozobot can be differentiated depending on the application. The overarching learning goal is to create a basic understanding of how automated, executable programmes work. The children learn that programming means that a human enters a sequence of specific commands into an informatics system, in this case the robot. These commands can be coded in different ways, e.g., with colours or through a visual programming language. In detail, the children thus learn about different ways of programming a system. In addition, they learn about algorithms and how they can develop them themselves. When using OzoBlockly, they also get to know the basic structures of programming languages. Since the Ozobot is ideally suited to being programmed by several children together, they also learn indirectly about goal-oriented communication via programmes and the challenges of teamwork.

**Reference to the goal dimensions of IT literacy**

Ozobot primarily addresses the "C2 Algorithms & Programming" *Content Domain* as well as the "P1 Modelling & Implementing" *Process Domain*. Due to its target group-oriented design, it can be used to develop algorithms independently and to implement them in different ways. Depending on the design of the discussions in the learning groups or further learning phases, features under "C4 Informatics Systems" and "C5 Computer Science and Society" can also be addressed.

In the *Process Domains*, another focus is on "P0 Interacting & Exploring", in that children can explore the programmable robot very freely. When playing in teams, Process Domain "P4 Communicating & Cooperating" is also applied.

## 3.5 LEGO WeDo 2.0

### Overview

LEGO WeDo 2.0[55] is an activity-based learning concept from LEGO that builds on the predecessor concept WeDo (see image). WeDo was developed especially for social studies and science in primary education ('Sachunterricht') and offers teaching materials in the form of projects that are already worked out and adapted to the curriculum. The kit for building these models contains various Lego bricks, wheels, sensors, e.g., for tilt and movement, a motor and a central control module, the so-called Smarthub. A connection between the Smarthub and the software on a PC, Mac, Android tablet or iPad can be established via Bluetooth. Among other things, this software offers a programming interface. With this, the created models can be programmed. WeDo and Wedo 2.0 have a connection via the visual programming language Scratch[56]. In addition, the software provides the material that guides through all the projects offered by LEGO: digital step-by-step building instructions, films, a digital learning diary and the aforementioned programming environment. Currently, however, the app seems to be more advanced than the software for the computer.

The WeDo concept introduces children to working and researching on the computer. Along the way, they learn and realise how many informatics systems and programmable devices there are in their own world. They learn on a small scale how devices are programmed and that various tasks can be solved with these devices and through programming.

### Target Group

According to the manufacturer and our own experience, LEGO WeDo 2.0 is suitable for primary school children from grade 2.

### User-friendliness

Even though LEGO WeDo 2.0 is very complex compared to many other systems presented here, it is very user-friendly due to numerous materials and the supporting software.

---

55  https://education.lego.com/de-de/products/lego-education-wedo-2-0-set/45300#wedo-20-set
56  https://scratch.mit.edu/wedo

**Previous experience and knowledge required of the learning supporter**

Learning supporters should have familiarised themselves with the LEGO WeDo construction kit and the model they are going to use in order to be able to support the children in case of questions or problems. Specific computer science background is not necessary when using Lego software. However, they should have familiarised themselves with the programming language for LEGO WeDo 2.0. If the language Scratch is to be used instead of LEGO's own programming language, learning supporters should have the necessary previous knowledge.

**Previous experience and knowledge required of the child**

Children do not require specific previous experience or knowledge. Many children already know LEGO from their own playroom. Building instructions for the suggested models are provided by the software.

**Learning objectives**

The IT-specific learning objectives in the context of WeDo relate, among other things, to the construction of an exemplary IT system. Children learn how sensors and actuators can interact in conjunction with motors. In connection with the programming, they also acquire a basic understanding of programming and fundamental programming concepts.

Using various examples, children gradually learn about different programmable systems and can thus relate to systems in their own life world.

Since children can work together, they also indirectly learn about goal-oriented communication via programmes and the challenges of teamwork.

Last but not least, children learn how to use an IT system.

**Reference to the goal dimensions of IT literacy**

LEGO WeDo 2.0 primarily addresses the *Content Domains* "C4 Informatics Systems" and "C2 Algorithms & Programming". The target group-oriented design enables primary-school children to design their own IT system and develop algorithms for this system in the context of problem solving (*Process Domain* "P1 Modelling & Implementing").

Depending on the design of the discussions in the learning groups and different learning phases, aspects under "C5 Computer Science and Society" can also be addressed.

In the other *Process Domains*, the focus here is on "P0 Interacting & Exploring", by allowing the children to create and explore their own IT system. In this context, the children can apply or develop interaction skills when dealing with this system. When playing in teams, Process Domain "P4 Communicating & Cooperating" is also applied.

## 3.6 Dash & Dot from Wonder Workshop

**Overview**

Dash & Dot[57] are a two-piece turquoise robot duo that look like a classic toy at first glance. Visually, they have a very child-friendly, robust design (also suitable for use in early child-care centres). The two robots can be controlled and programmed via various apps. Dash, the larger robot, can move, move its head, talk, light up and respond to sounds and obstacles. Its little brother Dot is immobile and mainly serves as a communication partner for Dash. In addition, numerous add-on sets are available (e.g., an adapter for other Lego extensions or mobile phone holder), leaving hardly any limits to the creative possibilities of use. With the various apps, children of pre-school and primary school age can explore the robot independently. Children can control the "Go" app manually and operate it intuitively. Other apps (such as "Wonder") facilitate programming with different approaches. The popular graphical programming environment Blockly can also be used.

**Target Group**

Depending on the app, even small children of pre-primary age can control Dash & Dot robots as easily as with a remote-control. Other apps additionally allow a deeper introduction to programming, so that children in primary school and even junior secondary school can solve challenging tasks with the robots or even design them themselves. According

---

57  *https://www.makewonder.de*

to the manufacturer's website, Dash & Dot are already used in more than 7,000 primary schools worldwide.

**User-friendliness**

Dash & Dot robots can be unpacked, charged and are ready to use immediately. As it is a very robust complete system, it is extremely user-friendly. Other user-friendly features include the very professional and child-friendly apps, although these only work on larger (newer) smartphones and tablets. These apps can be downloaded free of charge for Android and iOS.

**Previous experience and knowledge required of the learning supporter**

For the simple use of a remote-control app, learning supporters do not need to have any previous computer knowledge, they only need to be able to operate the tablet (or smartphone). Basic programming skills are required to programme the robot independently, although the apps are designed to be child-friendly so that even inexperienced learning supporters can quickly find their way around.

**Previous experience and knowledge required of the child**

Children should be familiar with using a smartphone or tablet, or time should be set aside to explore this medium. Beyond that, no previous knowledge is absolutely necessary. The system can also be explored via the remote-control app and further developed in subsequent steps with simple programmes.

**Learning objectives**

The overarching learning goal is a basic understanding of automatically running programmes and the communication of two robots with each other. Children learn how informatics systems (robots) can respond to each other with the help of sensors and thus communicate. Depending on the didactic design, children expand their problem-solving skills, develop their own algorithms and even formulate challenges themselves.

Children can independently explore possible components (e.g., sensors, motors) of an IT system with Dash & Dot, whereby this IT system cannot be changed.

**Reference to the goal dimensions of IT literacy**

The Dash & Dot robot couple primarily addresses the *Content Domain* "C2 Algorithms & Programming". Its target group-oriented design and the easy-to-use steps enable even children from pre-primary age to engage with the Content Domains "C4 Informatics Systems" and "C5 Computer Science and Society".

In the *Process Domains*, the focus at the beginning of use is clearly on "P0 Interacting & Exploring", as children can explore Dash & Dot very freely at different levels via apps. In further activities, the area "P1 Modelling & Implementing" can also be covered, whereby learning supporters can aim at modelling outside the apps (e.g., through cards that describe the state of the robots, which can then be arranged according to the sequence in which the robots are to process the different states). Depending on the didactic concept, the other Process Domains can also be covered.

## 3.7 Scratch and ScratchJR

**Overview**

Scratch[58] and ScratchJR[59] (junior version of Scratch) are graphical programming environments. This means that children do not have to write a textual programme code but have to put puzzle pieces (code fragments) together to make them fit. The three key elements of the interface of both applications are a stage, the menus with the commands and the programming area.

On the stage, the objects (e.g., a cat) execute the programmes (e.g., move or speak). Children drag and drop the commands from the menu onto the programming surface and connect them accordingly. The resulting programme can be tested at any time. This way, children can independently evaluate their results and improve them according to their goals.

In Scratch, the puzzle elements contain textual commands (sometimes combined with symbols), e.g., turn 90 degrees to the right. In the ScratchJR app, the commands consist exclusively of symbols, and the scope of commands is also much more limited.

In addition to simple movement instructions, both variants offer the possibility of changing the objects optically (e.g., make them smaller), sounds can be played or several objects can interact with each other. In Scratch, objects can also respond to numerous sensors (e.g., when the objects bump against the edge of the surface). Basic programming constructs (repetitions, branching etc.) are also available. All in all, there are no limits to the children's imagination and creativity due to the extensive range of commands (especially in Scratch).

In addition to using it for programming objects in a micro-world, there are extensions for Scratch that also enable programming of informatics systems such as the micro-controller Raspberry Pi or LEGO WeDo.

---

*58 https://scratch.mit.edu*
*59 https://www.scratchjr.org*

**Target group**

The original version of Scratch is suitable for children from fourth grade upwards, but the developers are also targeting 8-16-year-olds. The ScratchJR app can already be used expediently and without problems for children of pre-primary age, as no reading skills are required. In this case, the developers recommend the age group 5 to 7 years.

**User-friendliness**

Scratch has been available since 2007 already. Originally, it was a programme that users could download and install free of charge. For a few years now, Scratch has been offered browser-based (as a website), which eliminates the installation effort. Children can programme scratch without having an account, but they have to register to save their own projects (also free of charge). If online use is not possible, an off-line version is still available for download. Scratch has been translated into countless languages and is therefore also completely available in German, which makes it easier to use in primary schools.

ScratchJR is a free app for tablets (Android, iPad) and can be used without text.

Both are high-quality applications and very good to use both technically and didactically.

**Previous experience and knowledge required of the learning supporter**

In general, learning supporters should have an understanding of basic programming constructs. They should also know what a command looks like, how to combine different commands and be familiar with and able to use the constructs of repetition (loop) and branching (e.g., if-query).

**Previous experience and knowledge required of the child**

Children do not necessarily need any previous knowledge. In preparation, it would be useful to discuss everyday processes (e.g., brushing teeth), so that children develop a basic understanding of how they themselves carry out different instructions one after the other and possibly even repeat them.

## Learning objectives

The objective of both the Scratch and Scratch JR versions is to provide an introduction to the world of programming. The focus is not on the syntax of programming languages but on their basic concepts and simple algorithms. Children learn to programme their own processes in a micro-world. To accomplish this, they apply the basic concepts of instruction, repetition, and branching (also depending on sensor values). Furthermore, the focus is on algorithms.

## Reference to the goal dimensions of IT literacy

Both Scratch versions focus on *Content Domain* "C2 Algorithms & Programming". Child-friendly programming environments can help to plan and implement algorithms independently. Features under "C4 Informatics Systems" and "C5 Computer Science and Society" can also be addressed along the way.

In the *Process Domains*, the focus is on "P0 Apply & Explore", as well as "P1 Modelling & Implementing". Applying and exploring refers to the informatics systems tablet or laptop, as well as to working with a complex programme window with several areas. Because the game should definitely be presented at the end of the learning unit, the area "P4 Communicate (&Cooperate)" is also extended.

## 3.8 Makey Makey from JoyLabzLLC

### Overview

Makey Makey[60] was developed by Jay Silver and Eric Rosenbaum at MIT. It is a circuit board that can be used to find countless creative ways to turn everyday things into touch-sensitive surfaces and use them to interact with the computer. For example, an apple, a banana or a piece of modelling clay can be turned into a left mouse button.

Visually, Makey Makey looks like a game console control. It has 6 standard inputs: arrow keys for up, down, right and left, as well as the space bar and the left mouse button. 12 additional options enable, for example, full mouse control.

In addition to the circuit board, the Makey Makey set includes crocodile clips, jumper cables and a USB cable.

A simple control system can be built, e.g., with an apple. Makey Makey is connected to the computer via a USB cable. Two cables with crocodile clips are connected to Makey Makey; one for earthing, one, e.g., for the mouse click. The ends of the cables are connected to e.g., an apple and to the person. As soon as the person touches the apple, the electric circuit is closed and the corresponding

---

60 *https://scratch.mit.edu*

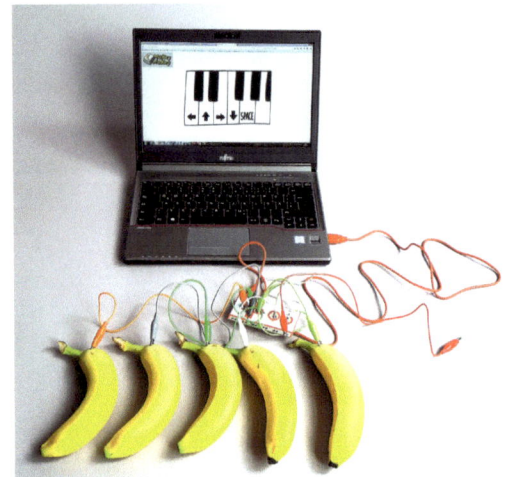

command, in this case a mouse click, is executed. In this way, various objects can become interactive devices. So, why not control the computer game with fruit for a change?

In connection with programming environments such as Scratch, further IT features can be discussed.

**Target group**

The target group for Makey Makey ranges from children aged 8 to adults.

**User-friendliness**

Immediately after setting up Makey Makey as an input device on the computer (this is done the same way as with a conventional mouse or keyboard), exploration with new interaction objects can start. After a short introduction to the handling, Makey Makey is very user-friendly.

**Previous experience and knowledge required of the learning supporter**

Learning supporters need to be able to set up Makey Makey as an input device in advance. They should also be familiar with the functioning of Makey Makey in order to be able to assist as contact person in case of questions or difficulties. If Makey Makey is also to be used in conjunction with Scratch, the relevant previous experience and knowledge of Scratch are also required.

**Previous experience and knowledge required of the child**

Children do not need any previous experience. However, knowledge of simple circuits is helpful to understand how it works.

**Learning objectives**

In terms of IT education, with Makey Makey children learn about interactions with different programs in an experimental way. They also learn how many different objects can be used as interaction devices. In addition to the aspect of input or interaction devices, a connection can also be established to the topic of electric circuits and conductivity of objects.

In conjunction with Scratch, Scratch-related learning objectives can also be achieved (see above).

Makey Makey is ideally suited for collaborative work, so that the corresponding competencies can be promoted on a cooperative level.

**Reference to the goal dimensions of IT literacy**

On the level of goal dimensions of IT education, Makey Makey primarily covers *Process Domain* "P0 Interacting & Exploring" in connection with *Content Domain* "C4 Informatics System". The focus of the application with Makey Makey is on exploring different ways of interacting with the computer or programmes on the computer. When working in teams, *Process Domain* "P4 Communicating & Cooperating" is also applied.

When using Makey Makey in combination with Scratch, the content and Process Domains listed there also come into play.

## 3.9 LEGO Mindstorms (NXT & EV3)

**Overview**

LEGO Mindstorms robots[61] are the most complex in the range of informatics systems introduced here. They can be considered the big sister of LEGO WeDo and have been conquering school and extramural learning spaces since 1998 (in different versions). The current version, EV3, entered the German market in September 2013, but the predecessor, NXT, is nowadays also still widely used.

Basically, the learning objectives of LEGO Mindstorms are analogous to those of WeDo, both in terms of building and programming the robot. LEGO Mindstorms is accordingly more variable and thus more complex in both areas (hard- and software). EV3 consists of a total of 600 individual parts, whereas the WeDo requires only 280 components. The decisive factors here are the three (instead of only one) motors and, depending on the version, up to four different sensors (colour, infra-red, gyro, touch sensor or microphone). In this way, children can build their own robots very creatively in addition to the given models.

Educators and learners also have many options when it comes to programming. Similar to WeDo, the robots can be programmed graphically with NXG software (for NXT), or EV3 programming software (for PCs and tablets). More diverse commands are available here, which can also be applied in a more differentiated way (e.g., exact distance values can be processed with the help of the ultrasonic sensor). Furthermore, LEGO Mindstorms can be programmed with numerous textual programming languages (e.g., Java, C++).

**Target group**

Lego Mindstorms are primarily designed for secondary-school children and are sometimes even used at universities for introductory courses in programming

---

61 *https://www.lego.com/de-de/mindstorms*

(then mostly in Java or C). They can also be applied for children at the end of primary school or in small groups with high supervision ratio. The recommended age for the current EV3 model is 10 years.

**User-friendliness**

Lego Mindstorms are designed for long-term use with school children and are therefore sturdily built. The software is also especially designed for programming beginners and is easy to use, although previous knowledge of using classic software (opening, saving, etc. of files) is required.

A handicap that often cannot be overcome when using Lego Mindstorms is the high price of currently €280 for the EV3, which includes the programmable brick, three interactive servo motors, three sensors (colour, touch, infra-red sensor with remote control) and corresponding Lego parts for building the proposed models.

**Previous experience and knowledge required of the learning supporter**

For the use of Lego Mindstorms, learning supporters should have a good basic knowledge of programming. In both the graphical and textual versions, programming constructs such as loops or conditions are applied very soon. When using textual programming languages, learning supporters must also have a good command of the syntax, otherwise it is almost impossible to trace errors in the learners' programmes.

**Previous experience and knowledge required of the child**

In principle, basic knowledge of classical software is sufficient (especially when using graphical programming languages), as the learners can get to know all programming constructs as well as the syntax of the programming language during the learning unit with Lego Mindstorms. However, especially when using textual programming language, it is recommended that children have already acquired basic skills of programming via another IT system in order to facilitate the introduction or transition to Lego Mindstorms.

**Learning objectives**

With Lego Mindstorms, learning objectives can be achieved both in the area of robot building (technology) and in the area of programming (computer science).

In the area of robot building, technical aspects can be addressed, such as the transmission through gears of different sizes or cornering through different speeds on both drive axes.

For programming, there are no limits in the area of imperative programming. School children can acquire all important basic programming constructs with the help of the robot. Another special focus is on dealing with sensor values.

**Reference to the goal dimensions of IT literacy**

This IT system is oriented towards Content Domains "C4 Informatics Systems", as well as "C2 Algorithms & Programming" and, depending on the design of the learning unit, also "C5 Computer Science and Society". As with Lego WeDo, learners can design (construct and programme) their own IT system with Lego Mindstorms. The extensive possibilities also facilitate implementation of more complex algorithms (e.g., automatic following a black line or finding the way out of an unknown maze). The extent to which social effects of robotics are addressed in addition, depends primarily on the design of the learning unit, but can be excellently integrated for the topics of robotics. Due to the numerous sensors, topic "C1 Information & Data" can also be addressed in a learning unit with Lego Mindstorms. All in all, the same goal dimensions as with LEGO We Do are aimed for at a higher level.

In the *Process Domains*, LEGO Mindstorms focuses mainly on "P1 Modelling & Implementing", with the emphasis on implementation in most applications. Pure exploration is much more difficult here than in LEGO WeDo or requires a higher level of guidance. The extent to which the areas "P2 Reason & Evaluate", "P4 Communicating & Cooperating" and "P5 Represent & Interpret" are additionally deepened depends on the design of the learning unit. Children and teenagers often work in pairs with the robot or have to agree as a team on an algorithm to solve a problem (e.g., the way out of the maze), which means that they also acquire and develop competencies in these Process Domains.

## 3.10 Arduino Microcontroller with ArduBlock

**Overview**

ArduBlock[62] is – analogous to Scratch or App Inventor – a graphical programming environment for use on the computer. The special feature of ArduBlock is that in the micro-world, objects are not controlled on the screen, but an Arduino Microcontroller[63] and actuators (e.g., motors) and sensors (e.g., a light sensor) connect-

---

62 *http://blog.ardublock.com*
63 *https://www.arduino.cc*

ed to it, are programmed. The same combinations can also be realised with other micro-controllers (such as Raspberry Pi) and also other programming environments (like Scratch). The Arduino Microcontroller and the various electrical components that can be connected to it represent the most flexible of the informatics systems described here and can be expanded as desired.

Little Robot Friends from Aesthetec Studio is a manageable kit that can also be assembled (soldered) by primary-school learners. They consist of only a few components, but thanks to the microphone, loudspeaker and LEDs, they offer diverse and creative programming possibilities. The communication/networking of several Little Robot Friends with each other is particularly exciting.

The Little Bits from littleBits Electronics Inc. are an alternative to this. With the basic set, in particular skills in the field of electronics (physics, electrical engineering) can be acquired by plugging electrical components together in such a way that, for example, an LED or a signal transmitter can be controlled via a button. These circuits can be programmed with the help of an Arduino add-on set. The Arduino can be programmed via ArduBlock, for example.

**Target group**

When ArduBlock software is combined with the Arduino, the target group depends greatly on the choice of components and complexity of the tasks. Initial experiments with a few LEDs and a motor (especially with the child-friendly components of Little Bits) can already be carried out at primary-school age. The actual goal of independently planning and implementing one's own projects with electronic components and their programming is more in line with the competence level of learners in the junior secondary grades (from grade 7 or 8).

**User-friendliness**

In contrast to the other examples, Arduino software with the ArduBlock add-on must be installed. The disadvantage here is the additional work that can only be done by external administrators, depending on the administration system of the institution. At the same time, offline software has the advantage that it runs stably and independently of the Internet connection and can also be used outside WiFi coverage.

In addition, ArduBlock is similar to Scratch in terms of user-friendliness, so that learners have a quick success experience and are not burdened by syntax mistakes of a textual programming language.

The associated hardware is very diverse and ranges from child-friendly LittleBits to classic sensors and actuators from electronics stores.

**Previous experience and knowledge required of the learning supporter**

This learning concept requires certain prior knowledge of both electrical engineering/physics and computer science. Learning supporters must be familiar with the physical basics of electrical circuits, know the function of the components used and be able to convey this in a target group-oriented way. They must also have sound programming skills (with a focus on the specifics of micro-controllers). All in all, this approach places the highest level of competence on the part of the learning supporters compared to the other approaches.

**Previous experience and knowledge required of the child**

A learning concept with Arduino is an exciting combination of technology and computer science and learners should have prior knowledge in both areas, so that they are not overwhelmed. Basic knowledge of physics (e.g., about electrical circuits) should be assumed as well as simple algorithms (e.g., how a traffic light system works) that the children can understand and reproduce on their own.

**Learning objectives**

By combining the programming environment (ArduBlock or Arduino) and the Arduino Microcontroller, comprehensive STEM skills can be covered. The focus here is on technical and IT learning objectives, as the functioning of various electrical components must be understood and combined expediently in the circuit, before the micro-controller that controls this circuit, can be programmed. Because of different options in the complexity of the components and the algorithms to be developed, the learning objectives are also very variable.

**Reference to the goal dimensions of IT literacy**

With the concept of combining hardware and software, almost all *Content* and *Process Domains* can be covered, although the distribution across the different areas depends on the implementation of the learning unit. For example, Content Domain "C3 Languages & Automata" requires an explicit focus on automatic machine-controlled programming.

## 4 Recommendations

We would like to conclude with a recommendation on how the informatics systems analysed above can be usefully applied with the target group of children in childcare centres or primary schools. As different as children, schools, childcare centres and learners or educators are, so are the corresponding recommendations.

The 10 informatics systems analysed above have different emphases on which goal dimensions of computer science education they aim at and on which prior knowledge they build, both on the part of the teachers or educators and on the part of the children. All in all, however, all systems have in common that they primarily address the Content Domains "C2 Algorithms & Programming" and "C4 Informatics Systems" and focus on the Process Domains "P0 Applying & Exploring" and "P1 Modelling & Implementing". For this purpose, the 10 systems were divided into three levels of complexity: low, medium and higher.

"*Low complexity systems*" include the following systems:

- Cubetto
- Beebot
- Kibo
- Ozobot without visual language

These systems can generally be used autonomously, i.e., no other technical system (computer, tablet or similar) is needed. All four are very suitable for use in child-care centres or for initial contact in primary schools and mainly address Process Domain "P0 Applying and Exploring". Cubetto, in particular, requires almost no instructions, since the wooden blocks specify the direction of movement and the robot processes the commands it receives. As such, this system can be used by any teacher or educator regardless of their computer science background. However, if the additional method (green brick) is to be used, a basic understanding of this programming construct is required. When using the BeeBot, the children must be able to memorise their programme (input via the keys), or the teacher or educator must support them, for example, by drawing the commands on paper externally. The two other systems, Kibo and Ozobot, increase the complexity through the use of sensors or a greater range of programming options.

All in all, these four systems do not present a major barrier to getting started and require only a low level of computer science competence on the part of the teacher or educator and no prior knowledge on the part of the children. Cubetto and Beebot are suitable for short-term use because they have hardly any barrier to getting started, but they too reach the limits of their functions after a few hours. Kibo and Ozobot are more suitable if long-term engagement with the system is planned.

After a first contact or also when starting with older children, a system with medium complexity is recommended.

The level of *"systems with medium complexity"* includes:

- ScratchJr
- Dash & Dot
- WeDo
- Makey Makey
- Ozobot with visual language (the special feature here is that this system can cover the entire range of complexity)

To use these systems, the teacher or educator needs further technical devices in the form of tablets (smartphones) or computers. This directly requires a different level of training on the part of the teachers or educators, as they have to support the children in using a classical system, i.e., one that is not specially designed for young children. Systems that are controlled via an app on a tablet are best suited for child-care centres, as the operation of a tablet via touch gestures is much more intuitive than that of a computer (or laptop) with a keyboard and mouse. Thus Dash & Dot and ScratchJR are appropriate examples here. One argument in favour of using the WeDo and Makey Makey systems is that the system itself can also be customised. Therefore, the Content Domain "C4 Informatics Systems" can be dealt with at a different level. Before children use these systems, they should have a general knowledge of computers (or laptops), i.e., they should know how to open and save files.

Since a programming language is used (in all the examples in this category), teachers or educators need some initial or even advanced computer science knowledge in programming, depending on the system. The teacher or educator must be familiar with basic programming constructs (e.g., loops, variables and branches) before using the systems.

Systems in the third category are primarily suitable for deepening knowledge or for use with older learning groups in secondary schools or small groups with a high supervision ratio, e.g., in extra-mural settings or extra-curricular working groups. These systems are not recommended for child-care centres.

"*Systems with higher complexity*" include:

- Scratch
- Lego Mindstorms
- Arduino Micro-controller with ArduBlock

These systems use a complete (graphical or textual) programming language. Even though in the case of a graphical programming language it is not necessary to learn the syntax (puzzle blocks instead of text input), the teacher or educator needs a higher level of expertise, both in terms of programming and computer systems. Since the systems in this category have considerably more extensive possibilities due to their higher complexity, the various defined goals can be dealt with in the long term and in greater depth. In particular, the domain "C2 Algorithms and Programming" can be dealt with comprehensively by means of these systems.

# 5 Conclusion

The analysis of the informatics systems presented here leads to the recommendation to start with a low-complexity system in child-care centres and primary schools in order to give both the teachers or educators and the children confidence to deal with informatics systems through a low entry hurdle and to consolidate the first important basics in dealing with these systems. Subsequently, depending on the age structure and previous knowledge of the children and teachers or educators, as well as the equipment of the learning venues, informatics systems with medium and/or higher complexity can and should be used to deepen the competencies in a spiral fashion. The focus of the competence acquisition with the informatics systems presented here relates to Content Domains "C2 Algorithms & Programming" and "C4 Informatics Systems", as well as Process Domains "P0 Applying & Exploring" and "P1 Modelling & Implementing". Of course, other Content and Process Domains can also be covered. However, it is not absolutely necessary to use the informatics systems presented here.

Informatics systems thus represent one of many ways in which computer science content can be taught in early education. Other possibilities are, for example, approaches that do without any Informatics Systems, so-called unplugged approaches. They are also suited to early computer science education.

Computer science education is necessary to gain a basic understanding of informatics systems, which play an increasingly important role in our everyday lives. In order to participate in today's digitalised world, it is essential that all children have the opportunity of early computer science education.

Even though there are different approaches to teaching computer science, we consider the sustained engagement with informatics systems aimed at different defined goals (from pure exploration to the development of one's own algorithms) as indispensable. Although there are currently no studies that provide proof of this, we assume that children can only grow up to be active co-creators of our current and, above all, future digital as well as real world if informatics systems are also used in the context of computer science education. Otherwise, children often lack the application reference or the specific context of the concepts of computer science they have learned.

# Conclusion and Outlook – How the "Haus der kleinen Forscher" Foundation uses the Findings

"Haus der kleinen Forscher" Foundation

1. Recommendations from the Expert Reports as a Basis for the (further) Development of the Foundation's Substantive Offerings
2. Digital Education – A Chance for Good Early STEM Education for Sustainable Development
3. Scientific Monitoring and Evaluation of the Professional Development Workshops
4. Outlook – Organisational Development in Educational Institutions

# 1 Recommendations from the Expert Reports as a Basis for the (further) Development of the Foundation's Substantive Offerings

All offerings of the "Haus der kleinen Forscher" Foundation are based on professionally sound goal dimensions for children as well as early childhood educators and primary school teachers in the respective disciplines. They serve as a guideline for the Foundation's programme content and specify which goals are to be achieved through specific offerings by the Foundation. Furthermore, the respective model of the goal dimensions provides the theoretical basis for the scientific monitoring and its empirical verification.

So far, the goal dimensions of science education in elementary and primary education (cf. Volume 5 of this series, "Haus der kleinen Forscher" Foundation, 2018), early technical education (cf. Volume 7, Stiftung Haus der kleinen Forscher, 2015) and early mathematical education (cf. Volume 8, Stiftung Haus der kleinen Forscher, 2017) have been developed and published. The present volume now contains the goal dimensions of computer science education in the elementary and primary sector and the resulting recommendations for the development of the Foundation's offerings in the subject area of "computer science".

The following describes how the "Haus der kleinen Forscher" Foundation takes up the recommendations offered by experts and implements them in order to develop its offerings in the field of computer science education for three- to ten-year-old children and the accompanying early childhood educators and primary school teachers at early childhood education and care centres, after-school centres and primary schools. Special attention was paid to ensuring practical relevance and strengthening educators in their role as learning supporters in the implementation of computer science education.

The Foundation pursues the following goal dimensions of computer science education at the level of the children (cf. recommendations Chapter 3 and Figure 25):

- Motivation, interest and self-efficacy when dealing with computer science
- Computer science process domains
- Computer science content domains

At the level of early childhood educators and primary school teachers, the Foundation focuses on the following goal dimensions recommended by Bergner et al. (cf. recommendations Chapter 4 and Figure 34):

- Motivation, interest and self-efficacy in computer science education
- Computer science process domains
- Computer science content domains
- Computer science didactic competencies
- Attitudes, mindsets and understanding of roles with regard to the design of computer science education

All of the Foundation's content formats aim to strengthen the development of children of pre-primary and primary school age in the relevant goal dimensions. Most of the Foundation's offerings initially support early childhood educators and primary school teachers who then act as supporters for children at the educational institutions as they come to grips with the subject matter, promoting children's learning and development processes. In doing so, the Foundation offers a practical approach that enables educators to expand their knowledge and competencies and draw on their expertise in their day-to-day interactions with the children.

The aim of early computer science education is to enable children to gain basic experience of computer science and to develop or promote a fundamental understanding of the subject. The individual goal dimensions and their concrete implementation in the Foundation's offerings for early childhood educators, primary school teachers and children are described in detail below.

## 1.1 Motivation, Interest and Self-Efficacy when Dealing with Computer Science

The first goal dimension recommended by Bergner et al. is "motivation, interest and self-efficacy in dealing with computer science". It applies to children and teachers alike and underlines the importance of enthusiasm, curiosity and interest as an essential to a positive approach to computer science. These include:

- interest in computer science (systems)
- motivation in dealing with computer science issues
- self-efficacy expectation in dealing with computer science
- for educators, additional motivation as well as self-efficacy expectations with regard to the design of computer science education

In terms of the motivational and emotional aspects, it cannot be ruled out that there will be a discrepancy between the early childhood educators and primary school teachers, on the one hand, and the children, on the other, in the field of computer science. Children are usually very motivated and interested, at least when it comes to using digital devices. Previously, the main influences will have been their own parental home, whereby children from education-oriented families tend to be more often confronted with reservations and fears regarding the use of such devices (Chaudron, 2015; cf. also Bergner et al. in this volume, section 2.1.1). Similar reservations can also arise among educators when it comes to the use of digital devices at the institutions and addressing computer science topics and issues. Insufficient skills or a lack of competence in the field of computer science education can lead to insecurities in dealing with the topic and therefore to avoidance, too.

However, according to a representative survey conducted by the "Haus der kleinen Forscher" Foundation, 75% of the educators questioned were in favour of learning how to use digital devices responsibly at early childhood education and care centres (Stiftung Haus der kleinen Forscher, 2017b). The vast majority of educators therefore see the responsible use of digital devices quite positively.

Moreover, early childhood educators and primary school teachers serve as role models for the children, and this also applies to their experience of and attitude towards computer science. Here, the children benefit from an exemplary interest on the part of their educational attachment figure, who is someone they can consolidate regarding questions relating to computer science. An open, fear-free approach to computer science enables children to build up an interest in the subject and maintain it in the long term.

For this reason, the implementation of this goal dimension focuses on fostering an open and fear-free attitude towards computer science both in children and in early childhood educators and primary school teachers. Educators are to develop a sense of pleasure in designing computer science education in day-to-day educational life and be motivated to pursue computer science questions together with the children.

**Implementation of this goal dimension in the Foundation's offerings**

For the "Haus der kleinen Forscher" Foundation, the focus is on having fun and taking pleasure in exploring and understanding the world around us. By tackling day-to-day questions about nature and technology, the aim is to encourage children's curiosity, joy of learning and thinking. In the field of computer science, too, the aim is to allow them to sense their own competence and self-efficacy in their day-to-day lives in an increasingly digitalised world.

When implementing the goal dimension "motivation, interest and self-efficacy", it is important to introduce the children and the early childhood educators and primary school teachers to computer science issues and enable them to have positive basic experiences. One aim here is to arouse their fascination with the creative and problem-solving character of computer science and not to encourage a purely receptive approach to digital devices. In particular, the aim is for educators to discover that topics of computer science not only concern computers, smartphones, etc., but arise in many ways in their own day-to-day lives and those of the children. Developing winning strategies, executing action stages and keeping secrets are just a few situations in which it is possible to identify learning opportunities of computer science. Recognising such situations thus becomes easier, thereby reducing any reservations about computer science.

The professional development workshops designed by the Foundation consistently aim to give early childhood educators and primary school teachers a positive approach to the topic and enable them to develop an open, fear-free attitude. In the area of computer science education, the Foundation developed its first professional development workshop in 2017 on the topic "Discovering computer science – with and without computers". In 2021, the Foundation expanded its range of offerings within the PRIMA! project (funded by the Federal Ministry of Education and Research, 2020-2023) to include the blended learning professional development workshop "Computer science education in primary school teaching". This is a professional development programme explicitly aimed at primary school teachers.

In order to counteract possible reservations towards computer science education in the elementary and primary sector and to boost enthusiasm and interest in the subject matter, the Foundation primarily pursues access without the use of digital devices ("unplugged"; cf. Bergner et al. in this volume, section 2.3.1) when designing its computer science offerings. This also ensures that the implementation of computer science education does not depend on the financial resources available to educational institutions. The key basics of computer science can be explored with paper and pencil, using everyday materials or by means of pure physical engagement. For this reason, the professional development workshops on computer science education are designed in such a way that institutions do not need digital devices to create their own learning opportunities. Even without using a computer, together with the children it can be explored, for example, how computers sort numbers, how messages are encrypted or why digital pictures are made of pixels. On the one hand, this unplugged approach is well suited to exploration- and inquiry-based learning with young children, while on the other hand, it reduces inhibition thresholds among early childhood educators and primary school teachers. As with the Foundation's other educational areas, the promotion of computer science competencies can happen on a hands-on, day-to-day basis.

However, the Foundation also uses two other approaches to computer science, namely software-based entry (cf. Bergner et al., section 2.3.2) and physical experiential access to programming (robotics; cf. Bergner et al., section 2.3.3). The expert group assumes that these approaches are very motivating, especially for children, because they receive direct feedback on their work. For the purpose of software-based access, there are graphic programming environments with low levels of complexity that can be used and operated even without reading skills or much previous experience on the part of the children. The children learn to use programming as a creative tool and create their own "product" (e.g., using the programming language Scratch to control the movements of figures on the display by putting together graphic puzzle pieces). Child-friendly robot systems enable programming in the physically tangible world. Children can interact directly with such informatics systems and obtain immediate feedback in the course of the interaction (e.g., they can have the robot Cubetto move around a chair by operating the programming board). Regardless of the approach chosen, computer science education strengthens children in their problem-solving skills, and they experience self-efficacy by creating their own products.

In the professional development workshops, participants receive concrete practical ideas that also illustrate how computer science relates to everyday life. Likewise, numerous illustrative ideas that facilitate access are included in the pedagogical resources, such as the exploration cards, and the thematic brochure "Discovering computer science – with and without computers" (Stiftung Haus der kleinen Forscher, 2017c). The aim here is for early childhood educators and primary school teachers to learn to identify computer science content in day-to-day situations. This practical relevance strengthens both their motivation and their self-confidence. The practical examples are also chosen in such a way that they are challenging for both children and learning supporters but are also within their capabilities, thereby awakening a sense of satisfaction in tackling computer science questions or problems (cf. Bergner et al., section 3.2.1).

With the aim of integrating computer science education in the classroom for children of primary school age, the Foundation has developed the three-month blended learning professional development workshop "Computer science education in primary school teaching". The multi-section professional development programme consists of alternating and interrelated online and face-to-face phases which enable teachers to apply the knowledge they have acquired in parallel in the classroom, sharing their experience soon afterwards and supporting each other in their further development. After an introductory webinar, the educators first ask about the children's prior knowledge and interests. The Foundation provides teachers with an interview guide for this purpose. They use the children's ideas as a starting point for designing their lessons. This pupil-centred approach increases children's motivation by building on their interests and prior knowledge. In the subsequent practical webinar, participating teachers reported that they had observed children developing their problem-solving skills through exploration- and inquiry-based learning, which in turn has an impact on children's self-efficacy. Within the blended learning professional development workshop, the teachers receive lots of practical ideas for the classroom aimed at promoting their motivation as well as their self-efficacy expectations with regard to the design of computer science education.

## 1.2 Computer Science Process Domains

The goal dimension "Computer science process domains" is described by Bergner et al. as a significant domain of competence both at the level of the children and at the level of the early childhood educators and primary school teachers. According to Bergner, the process domains describe the way in which the children are to tackle subject matter (section 3.3). In deriving individual process-related competencies, the expert group was guided by the standards for the junior secondary level proposed by the Gesellschaft für Informatik (German Informatics Society, GI – Gesellschaft für Informatik e.V., 2008) as well as existing international curricula of computer science education. These educational standards were expanded to include the process domain "Interacting and Exploring" in order to emphasise the importance of a hands-on exploratory handling approach to informatics systems in the pre-primary and primary school sector (cf. Bergner et al, section 2.5.2). The experts therefore recommend the following process domains:

- Interacting and Exploring
- Modelling and Implementing
- Reasoning and Evaluating
- Structuring and Interrelating
- Communicating and Cooperating
- Representing and Interpreting

The authors of the expert report emphasise that the individual processes are always to be developed and applied in connection with one or more content domains (see the following section 1.3). In principle, all process domains can be linked to all content domains. However, there are combinations that would appear to make more sense than others, especially with regard to the age of the children. In addition, numerous practical examples often also refer to several process and content domains. Thus, these can be addressed simultaneously in such cases, albeit to differing degrees.

In the expert report, computer science is described as a constructing science, i.e. the development of an informatics system is subject to a construction process (Bergner et al., in this volume, section 1.3). Cyclical models are used to develop new products or to adapt and further develop existing ones. According to the expert group, the focus of computer science education should be on processes of planning and designing. A design cycle for the construction of digital artefacts for this purpose was developed and proposed by Bergner et al (section 1.5.2). This design process could be preceded by the exploration of informatics systems, especially in early computer science education. This is also a cyclical process in which the individual components and the mechanisms of action of the system can be explored based on the functions and the intended use. An exploration cycle for this purpose was also proposed by the expert group (section 1.5.1). Both the design cycle and the exploration cycle underline the process-based character of computer science.

**Implementation of this goal dimension in the Foundation's offerings**

In the implementation of the goal dimension "Computer science process domains", the Foundation pursues the goal of familiarising both children and early childhood educators/primary school teachers with the process-based approach of computer science. They are to get to know and apply cyclical approaches. The focus here is on teaching general thinking and problem-solving skills, which in turn means that computer science education is able to contribute to the children's general education.

Practical ideas have been developed by the Foundation for all process domains suggested by the expert group. These are exemplified in the professional development workshops "Discovering computer science – with and without computers" and "Computer science education in primary school teaching". Educators can find further ideas for computer science education in day-to-day life with the children on the exploration cards and the thematic brochure contained in the accompanying package of materials (Stiftung Haus der kleinen Forscher, 2017d). The process domain "Interacting and Exploring" derived and supplemented in the expert report has more of a subordinate role to play in the Foundation's range of topics. This is due to the fact that the Foundation has decided to mainly develop practical ideas that do not require digital devices and can instead be implemented using everyday materials. Nevertheless, the Foundation also sees this process domain as an important competence and refers to the relevant reference to informatics systems in the exploration cards, which in some cases also includes exploring this system. So this process domain can also be implemented, depending on the technical resources available at the educational institution. In the course of piloting the professional development workshop "Computer science education in primary school teaching", the Foundation's specialists were able to observe that exploring and trying out informatics systems offers enormous learning potential for children, and that they should be given sufficient time for this in the classroom.

Furthermore, references to the process domain "Interacting and Exploring" can be found in the thematic brochure (cf. Figure 45), for example, in the chapter "Robots – from wonder to control". Suitable age-appropriate systems were selected and presented based on the expert recommendation by Bergner and Müller (in this volume). In these informatics systems, the actual programming is preceded by discovering and trying out the robot systems. As already described in the expert report, the individual ideas for practical implementation cannot always be clearly assigned to one process domain but apply to several areas at the same time.

**Figure 45.** Title page and table of contents of the thematic brochure "Discovering computer science – with and without computers" (Stiftung Haus der kleinen Forscher, 2017c)

In order to clarify process orientation in relation to computer science, the "Computer Science Cycle" (cf. Figure 46) was developed together with experts in the field – based on the "Inquiry Cycle" method[64] used in early science education (Stiftung Haus der kleinen Forscher, 2013, cf. also the Foundation's pedagogical approach in Stiftung Haus der kleinen Forscher, 2019), as well as the "Mathemat-

---

64 cf. Marquardt-Mau, B. (2004). The didactic concept of basic inquiry-based science education with children and the associated inquiry cycle model was developed by Prof. Dr Brunhilde Marquardt-Mau (2004) and adapted for the pedagogical approach of the "Haus der kleinen Forscher" Foundation.

ics Cycle" (Stiftung Haus der kleinen Forscher, 2016) and the "Technology Cycle" (Stiftung Haus der kleinen Forscher, 2018).

The "Computer Science Cycle" is based on the exploration and design cycle described in the expert report and adapts the individual phases to the needs of early childhood educators and primary school teachers in the elementary and primary sector. The aim of the cycle is to clarify and support the computer science approach. It is divided into six phases: (1) Formulate a question or need from a computer science perspective, (2) Describe a situation specifically, (3) Develop a model, (4) Apply the model, (5) Evaluate the result, and (6) Discuss the results and process. The "Computer Science Cycle" thus includes phases of concrete action and phases of documentation and reflection. In applying the cycle, it is possible to adopt both the perspective of exploring existing informatics systems and the perspective of designing a new product, as well as switching between the individual phases of these perspectives.

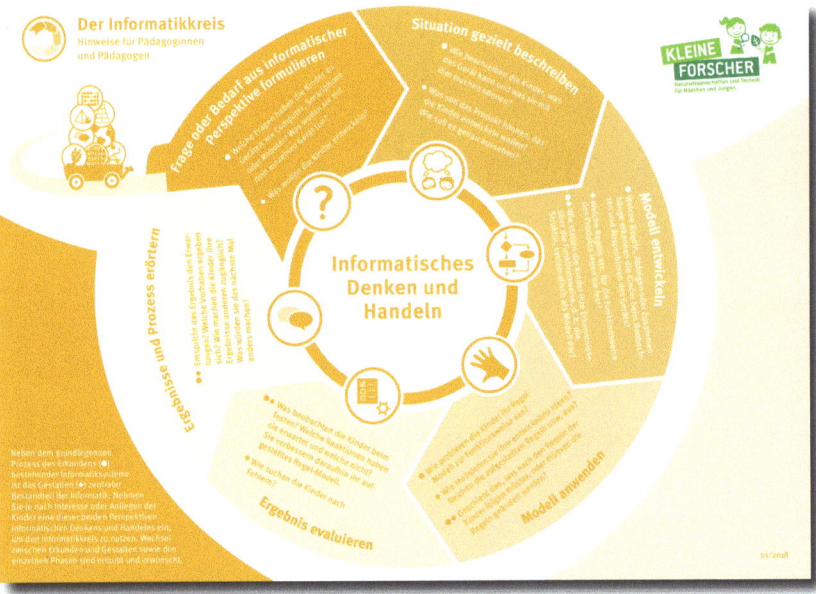

*Figure 46.* The "Computer Science Cycle" maps the process of the computer science approach and includes the perspectives of both exploring and designing

Early childhood educators and primary school teachers also have access to an online course on the "Computer Science Cycle" in which they learn to actively explore informatics systems with the children and develop their own applications together – in other words they get to shape their own digital environment. The course contains numerous suggestions for practical work and example videos of activities at early childhood education and care centres and primary schools.

In addition, teachers can find out more about the "Computer Science Cycle" in a mandatory consultation session offered as part of the further professional development workshop "Computer science education in primary school teaching".

## 1.3 Computer Science Content Domains

The goal dimension "Computer science content domains" is also prioritised by Bergner et al., both at the level of early childhood educators and primary school teachers and at the level of children. The content domains specify the computer science topics to be acquired by educators and children (section 3.3). As with the computer science process domains, the experts based the design of this competence area on the proposed standards for the junior secondary level as defined by the Gesellschaft für Informatik (GI – Gesellschaft für Informatik e.V., 2008). They distinguish between the following five content domains:

- Information and Data
- Algorithms and Programming
- Languages and Automata
- Informatics Systems
- Computer Science and Society

Orientation towards the GI standards for children of pre-primary and primary school age enables coherence across the entire educational chain, thereby establishing the basis for systematically developing computer science competencies in secondary schools by building on day-to-day computer science and play experience in the elementary and primary sectors.

As already mentioned, the content domains are always to be linked to one or more process domains, or else several content domains can be linked to one and the same process domain. According to the expert group, the following combinations of content and process domains are particularly suitable for elementary and primary education (cf. Bergner et al., in this volume, section 3.4.1):

- Modelling and Implementing of Algorithms and Programmes
- Interacting with and Exploring Informatics Systems
- Representing/Presenting and Interpreting Information and Data
- Reflecting on and Evaluating the Interrelationship of Computer Science and Society

In selecting these competence domains, the expert group focused on aspects such as relevance to day-to-day life, children's interest and contribution to general education, as well as subject-specific didactics and the psychology of learning and development. The prioritised combinations of content and process domains seek to promote general basic competencies and therefore general education, in addition to subject-specific didactic competencies in the area of computer science.

**Implementation of this goal dimension in the Foundation's offerings**

With the development of an offering for early computer science education, all four STEM subjects have now been represented in the Foundation's range of educational offerings since 2017. In contrast to the previous professional development topics in mathematics, sciences and technology, which were more differentiated in terms of content (e.g., "Exploring mathematics in space and form" vs. "Numbers, counting, calculating – exploring mathematics" with a focus on numbers and operations), all content domains suggested by the expert group for the first professional development workshop "Discovering computer science – with and without computers" and the associated practical ideas were taken into account in the Foundation's development of a programme for computer science education. As in the case of the process domains, individual ideas regarding practice can be assigned to several content domains. In developing these ideas, consideration was given both to practical implementation with children of pre-primary and primary school age and the combinations of content and process domains as prioritised by Bergner et al.

The professional development workshop "Discovering computer science – with and without computers" and the accompanying package of materials (Stiftung Haus der kleinen Forscher, 2017d) – with the thematic brochure, exploration cards (Figure 47) and hidden object picture (Figure 48) – support early childhood educators and primary school teachers in exploring the topic of computer science in day-to-day life with the children and in using this to promote the development of the children's computer science skills. In the professional development workshop, the reference to day-to-day informatics systems, such as traffic lights, smartphones and fully automatic washing machines, is clarified so as to make the world of computer science visible and tangible to early childhood educators and primary school teachers and therefore to children, too. In addition, educators learn about the content domains of computer science in the professional development workshop and explore the relevant practical ideas together. In addition to the in-person professional development workshop, the "Haus der kleinen Forscher" Foundation also offers an online course entitled "Discovering computer science" in which early childhood educators and primary school teachers also learn about the computer science content domains and receive suggestions for practice.

*Figure 47.* Front and back of the exploration card "What's it like to be a robot. Planning and controlling processes" (Stiftung Haus der kleinen Forscher, 2017d)

The professional development workshop "Computer science education in primary school teaching" focuses on the two essential content domains of "Data and Information" and "Programming and Robotics", since the aim is for children to gain basic experience primarily in these areas[65]. In the online module "Practical ideas regarding the basic concepts of computer education", teachers receive numerous different suggestions and ideas for designing computer science education in teaching practice within these two content domains.

For children of primary school age, the Foundation also developed exploration cards and the children's website www.meine-forscherwelt.de (currently "Fabio's Surfaces", "Ronja's Robots") to promote exploration- and inquiry-based learning.

*Figure 48.* Hidden object picture to discover and explore informatics systems in day-to-day life (as a picture card or poster, Stiftung Haus der kleinen Forscher, 2017d)

## 1.4 Computer Science Didactic Competencies

As a significant goal dimension at the level of early childhood educators and primary school teachers, Bergner et al. recommend "computer science didactic competencies". Together with the process and content-related computer science competencies, this competence domain forms the professional basis for effective

---

65 The Foundation developed the targeted basic ideas on computer science in cooperation with the Chair of Computer Science Education at TU Dresden (Prof. Bergner and Christin Nenner).

computer science education in the pre-primary and primary school sector. The subject didactic competencies include:

- Basic computer science didactic competencies (understanding of the goals of computer science education)
- Competence to plan computer science learning environments and learning situations
- Contextual didactic action competence in computer science
- Competence to diagnose and evaluate computer science learning situations
- Pedagogical-professional communication skills with participants

The aim is to strengthen the pedagogical action strategies used by early childhood educators and primary school teachers. An understanding of the goals of computer science education, different didactic methods and the design of effective learning environments is of great importance. In addition, it is particularly important at the elementary and primary level to be able to identify and use day-to-day and play situations with computer science content so as to create computer science learning opportunities. In doing so, early childhood educators and primary school teachers are to be able to assess the developmental level of individual children and adapt their support accordingly.

**Implementation of this goal dimension in the Foundation's offerings**

The Foundation's offerings aim to familiarise early childhood educators and primary school teachers with concrete pedagogical approaches which they can then use to support the children as they engage in their learning processes. They also become familiar with numerous practical ideas as a stimulus.

At the beginning of the in-person professional development workshop "Discovering computer science – with and without computers", the focus is on discovering computer science and its importance in day-to-day life. Early childhood educators and primary school teachers learn the definition of the term computer science and the goals of computer science education. The relevant thematic brochure (Stiftung Haus der kleinen Forscher, 2017c) also addresses the importance of computer science and computer science education and explains how to provide guidance and support in the latter. As the professional development workshop progresses, participants get to try out concrete methods in phases of practice and gain practical experience. In subsequent phases of reflection, participants discuss how what they have learned can be implemented in their practical work with children. In this way, the educators receive stimuli to integrate computer science learning opportunities in their day-to-day teaching. The alternation between practical experience and phases of reflection enables early childhood educators and primary school teachers to further develop and deepen their didactic competencies in computer science. In order to clarify computer science links in day-to-day life, practical examples are presented and applied in the training. In this way, educators are to learn to assess the computer science potential offered by play situations and day-to-day situations and become aware of how to make use of this. The expert group sees the professional competence to be able to recognise and design effective learning environments as a fundamental condition for the success of effective early computer science education (cf. Bergner et al., in this volume, section 4.4.2).

The blended-learning professional development workshop "Computer science education in primary school teaching", which is designed to last three months, includes introductory, practical and reflective webinars that build on each other, as well as self-study phases with online modules, interactive topic consultation sessions, group coaching and, above all, practical testing in the classroom. Once the participants have become acquainted with ways of thinking and acting in computer science in the introductory webinar, they deepen their understanding in the practical seminar based on practical ideas. In the online module "Practical ideas regarding the basic concepts of computer science education", teachers also receive numerous different suggestions and ideas for designing teaching practice in computer science education. For each teaching idea, there is a description of the time and material required for implementation as well as possible learning experiences for the pupils. Teachers can draw on the concrete suggestions to engage in exploration and inquiry with children and develop initial teaching ideas. For this purpose, they are also provided with templates for lesson documentation. In the online module "Tips for lesson planning", teachers receive further suggestions to help them plan their lesson and guide the exploration and inquiry process with their pupils. Once the teachers have tried out their practical ideas in class, they reflect on their experiences in the reflection webinar, where they can ask questions and engage in discussion. At the same time as they work through the online modules, teachers are provided with reflection portfolios to record their thoughts and ideas as they work through them. They can also draw on these portfolios to develop their teaching ideas.

In the professional development workshop, early childhood educators and primary school teachers learn about three approaches to computer science education. The practical ideas developed by the Foundation can mostly be implemented without digital devices in order to offer educators – as in the other topics covered by "Haus der kleinen Forscher" – the opportunity to work with everyday materials and to make it easier for them to get started in computer science education. For this reason, the focus of the workshop and the pedagogical resources is on access to the topic area without a computer. The early childhood educators and primary school teachers also get to know two other approaches to computer science education: a software-based introduction to programming and programming using age-appropriate robot systems. They are provided with information and suggestions for reflection on these different opportunities, also indicating the limits and challenges of the three approaches in their role as learning supporters as well as in terms of the children themselves. Depending on the preferences and resources of the educational institutions, educators can choose between these three approaches and use them for their practical work with children. The focus is always on exploration- and inquiry-based learning.

## 1.5 Attitudes, Mindsets and Understanding of Roles with Regard to the Design of Computer Science Education

The attitude of learning supporters towards computer science has an enormous impact on the computer science learning support provided at early childhood education and care centres, after-school centres and primary schools. For this reason, the authors of the expert report consider the goal dimension "Attitudes, mindsets and understanding of roles with regard to the design of computer science education" to be very important at the level of the early childhood educators and primary school teachers. This goal dimension includes:

- Beliefs about the nature of computer science
- Beliefs about the importance of computer science education at early childhood education and care centres, after-school centres and primary schools
- Beliefs about teaching and learning computer science
- Professional role and self-image

A positive attitude towards the subject of computer science is the foundation for successful engagement with computer science topics in day-to-day life with the children. The attitudes, beliefs and role expectations of early childhood educators and primary school teachers with regard to the design of computer science education are related to the goal dimension "Motivation, interest and self-efficacy (in computer science education)", as described in section 1.1. Fears, concerns and reservations of educators towards computer science issues and the use of digital devices or a lack of skills in this area can lead to a sense of insecurity in dealing with computer science topics and even to their avoidance. Interest in computer science and the self-efficacy of early childhood educators and primary school teachers with regard to computer science education are of great importance when it comes to developing an interest in computer science in children.

In contrast to other areas of education, computer science education in the elementary and primary sector still receives little attention. There is a lack of universally applicable standards, which makes the value of computer science education even more dependent on individual beliefs. One goal should therefore be for early childhood educators and primary school teachers to attach appropriate importance to early computer science education. According to Bergner et al., the relevant background knowledge and the development of subject didactic competencies among educators are key requirements for the development of a positive attitude towards computer science and for the support of children's learning processes in this subject area on a day-to-day basis.

**Implementation of this goal dimension in the Foundation's offerings**

The Foundation's offerings in the field of computer science aims to instil a positive basic attitude in early childhood educators and primary school teachers towards early computer science education. The numerous practical examples provided in the in-person and online training formats as well as in the accompanying materials illustrate the great relevance of computer science to everyday life, making it easier for educators to identify and make use of appropriate learning opportunities. The knowledge of the subject itself and subject-specific didactics acquired through the professional development workshops enables early childhood educators and primary school teachers to familiarise themselves with the topic and also to develop a positive attitude towards it themselves. Likewise, the Foundation's programme is geared towards helping early childhood educators and primary school teachers internalise the creative and problem-solving nature of computer science, which makes an important contribution to children's general education. Practical engagement with early computer science education aims to promote a positive attitude towards the subject, thereby also boosting the status of computer science as part of the canon of educational subjects.

## 2 Digital Education – A Chance for Good Early STEM Education for Sustainable Development

In establishing the subject-specific basis for early computer science education, the "Haus der kleinen Forscher" Foundation has sought to clarify the terms "digital education", "computer science education" and "media education", which are often blurred in public perception. The Foundation regards digital education as an umbrella term for media education and computer science education. In media education, the focus is on digital media as tools for producing one's own content and critically reflecting on the use, meaning and impact of these media. In computer science education, the focus is on understanding the basic concepts of automated information processing and applying these concepts to solve problems and understand information society. Dealing with informatics systems and digital tools with the aim of understanding and reflectively applying their basic principles can be an important aspect of computer science education. However, basic concepts of computer science education can also be discovered and understood without computer systems using non-digital means, thereby facilitating access – especially in the context of early education.

For this reason, the Foundation's professional development workshops in the field of early computer science education are designed in such a way that educational institutions can implement computer science education with or without the use of informatics systems, depending on their technical resources. Here, the Foundation promotes an age-appropriate, critical, creative and active use of digital media in the day-to-day teaching of educational institutions. Meaningful use of digital media at early childhood education and care centres and primary schools can promote early education, supporting the development of important future skills in an increasingly digitalised world.

With its position paper "Digital education – a chance to support good early STEM education for sustainable development", the Foundation provides orientation in the often critical discourse regarding the targeted use of digital media in early education

(Stiftung Haus der kleinen Forscher, 2021). The Foundation sees this as an opportunity to further develop the quality and impact of STEM education for sustainable development. Digital media form a fundamental part of a child's world and can play a key role in educational processes. Children should acquire the ability to understand digital media, use them responsibly, reflect on them critically and also apply them to explore the world. In addition, the Foundation is committed to ensuring that early childhood educators, primary school teachers and leaders have the opportunity to take online professional development workshops as well as in-person workshops, and that these enjoy equal recognition. For this reason, the Foundation considers digital media as being equal to other tools in the repertoire of child education and adult education and as an extension of the variety of methods in exploration- and inquiry-based learning.

The Foundation has formulated central theses in the paper that reflect its current position on STEM education for sustainable development in a world shaped by digitalisation:

1. STEM education for sustainable development empowers children for the future – also in a world shaped by digitalisation.
2. The didactic basis for the use of digital media, which is monitored by educators and teachers, supports children in investigating and exploring.
3. Digital media are equally important tools in the repertoire of good co-constructive learning facilitation.
4. Digital media promote continuous professional development support that is geared to the individual interests and needs of the learners.
5. STEM education for sustainable development with digital media requires conducive framework conditions in educational institutions.

For a meaningful and targeted use of digital media at educational institutions, early childhood educators and primary school teachers are needed who feel competent to support learning with digital media. With its qualification programmes for STEM education for sustainable development, the Foundation aims to encourage and enable educators and teachers to use digital media according to the pedagogical approach pursued by the educational initiative. In order to develop a new range of workshops focusing specifically on the use of digital media for exploration- and inquiry-based learning, the Foundation organised a forum in March 2020 to consult with cooperation partners – both academics and practitioners – and discuss the conditions for success for linking early STEM education and the use of digital media. The participants agreed that it was important to explore the creative potential of media with children and that the professional development workshops should focus on practical ideas for the use of such media in exploration and inquiry.

In the webinar "STEM goes digital", which the Foundation has offered since 2020, early childhood educators and primary school teachers learn how digital media can be applied in a worthwhile way to engage in exploration- and inquiry-based learning with children. Educators receive suggestions and stimuli on how to support children in pursuing their own questions and implementing ideas using digital media. In the webinar, as well as in the accompanying e-book (Stiftung Haus der kleinen Forscher, 2020a), educators receive background knowledge and practical ideas on the use of digital media in early STEM education for sustainable development and learn why it is worthwhile to use digital media, how this can work effectively and what to look out for as learning supporters. They are given suggestions for their pedagogical work, are able to draw inspiration from practical examples provided by different institutions and have the opportunity to compare notes and ask questions in the online forum.

The Foundation has also initiated a range of other projects to develop educational programmes in collaboration with educators. In the collaborative project "Digital Lab 2.0 – When Teachers Become Developers" (2019-2021, funded by the E.ON Foundation), the Foundation developed and tested the learning app "Potz Blitz! My Electricity Workshop" in cooperation with around 20 teachers for use in a teaching context. The app is intended to encourage teachers to make good use of digital applications in science lessons and support them in pursuing an interactive exploratory approach to the topics of electricity and energy with their pupils. A teaching guide describes the educational learning objectives of various subjects covered by the app and gives examples of how teachers can integrate the app into lessons in a didactically meaningful way. In addition, worksheets or handout templates are available for the preparation and implementation of their lesson.

With its project "Collaborative Concept Lab" ("Ko-Lab" for short, 2021-2022, funded by the Friede Springer Foundation), the Foundation is following up on the professional development programme "MINT goes digital" and the project "Digital Lab 2.0". Ko-Lab is developing a professional development workshop for primary school teachers on the topic of inquiry-based learning with digital media. Similar to "Digital Lab 2.0", the professional development concept is being evolved co-creatively with teachers as well as academics and practitioners. The aim is to support primary school teachers in designing their lessons in such a way that pupils engage in exploration and inquiry in relation to STEM topics with the help of digital media and develop competencies for a digitalised world. The resulting professional development elements are tested and adapted with the respective target groups (children, teachers, trainers). By the end of the project in autumn 2022, the aim is to have produced both a guide for the new teacher professional development workshop and a train-the-trainer manual for the qualification of multipliers.

## 3 Scientific Monitoring and Evaluation of the Professional Development Workshops

The Foundation has carried out various pilots and surveys in connection with its development of offerings of computer science education. The newly developed ideas and materials regarding implementation were first tried out at pilot institutions to test their practical suitability before being made available to all institutions. The professional development workshops were also tested and evaluated in practice with early childhood educators and primary school teachers in order to assess their impact and allow improvements to be made.

The professional development workshop "Discovering computer science – with and without computers" was evaluated in connection with the pilot project in cooperation with a Hamburg-based provider of early childhood education and care (Brünger, Franke-Wiekhorst, Griffiths, Günther & Radtke, 2019; Stiftung Haus der kleinen Forscher, 2020b). The participating early childhood educators and primary school teachers reported that their attitudes to computer science had changed as a result of attending the workshop. They perceive the importance of computer science in their own daily lives much more than before, which in turn provides them with multiple points of reference for early computer science education. The results also show that after participating in the professional development workshop, there is an increase in educators' motivation to continue engaging with the subject and they are looking forward to implementing computer science education with children. They also exhibit increased self-efficacy expectations in terms of implementing learning content in pedagogical practice, as well as higher levels of self-assessed knowledge and subject-specific didactic competence.

Trials of the professional development workshop "Computer science education in primary school teaching" with pilot teachers also showed that after attending the professional development workshop, teachers generally see a greater day-to-day relevance of computer science in their own day-to-day lives and in those of the children. They also have a higher level of subjective self-efficacy and confidence in planning and implementing computer science education in the classroom. In addition, they rate their subject knowledge and their subject-specific didactic competence in relation to computer science education more highly after attending the professional development workshop.

Nonetheless, transfer of training content to pedagogical practice remains partly subject to obstacles deriving from the educational institution as a whole, in particular the structural and cultural framework conditions at the institution (e.g., the pedagogical principles). During the piloting of the professional development workshop "Discovering computer science – with and without computers", respondents mentioned the support of the early childhood education and care centre leader and a good team structure as being important factors for successful implementation of computer science education at the early childhood education and care centre. In the opinion of educators, early childhood education and care centre leaders should be sensitised to the importance of educational content of computer science and contribute to structurally establishing the implementation of this material by means of internal communication measures. They said that this also influenced the entire early childhood education and care centre team, enabling the necessary time, personnel or financial resources to be made available to successfully implement computer science education on a day-to-day basis at the early childhood education and care centre (Brünger et al., 2019; Stiftung Haus der kleinen Forscher, 2020b).

In addition to the surveys in the pilot phases, feedback from the early childhood educators and primary school teachers participating in the professional development workshops is continuously collected and evaluated by means of monitoring measures (e.g., feedback forms, regular surveys). This gives the Foundation the opportunity to gather key insights regarding the implementation of the professional development programme in the field, also beyond the trial phase. As such, the Foundation's programme is not only theoretically well-founded, it also benefits from the practical experience of early childhood educators and primary school teachers. The materials are reviewed regularly and are revised and adapted as needed.

With the expansion of the Foundation's programme to include computer science, the Foundation now offers educational workshops in all four STEM domains. In contrast to the other areas of education, however, the Foundation has entered relatively uncharted territory in taking on computer science education in the elementary and primary sectors. For this reason, it is particularly important here to observe implementation in practical day-to-day teaching with the children, to gather insights about which concepts can be implemented with children of pre-primary and primary school age and where adjustments in the professional development programme are necessary. The expert group emphasises the importance of accompanying evaluation of the developed concepts "as an essential factor for the successful implementation of early computer science education in practice" (Bergner et al., in this volume, p. 235). For this reason, scientific impact research would also be desirable in the field of computer science education in order to gather more findings on the effects of computer science workshops among early childhood educators, primary school teachers and trainers, as well as on the impact at the level of the children themselves. To this end, the Foundation is in dialogue with academics with the aim of potentially initiating implementation and impact studies.

The Foundation will continue to maintain professional dialogue with external academics and practitioners to support the (further) development of workshop content in the field of computer science. The Foundation's concepts are regularly presented at specialist conferences and discussed with experts representing other institutions and practical initiatives in the field of computer science. A Scientific Advisory Board supports the Foundation on research issues and in ensuring that the Foundation's programmes are professionally sound. Expertise in computer science didactics has been represented on this Board since 2017, initially with Prof. Johannes Magenheim and since 2021 with Prof. Nadine Bergner. In addition, the results from the continuous evaluation and quality monitoring of the various Foundation workshops are incorporated into further developments in this area. Regular reflection and impact-oriented (further) development of the Foundation's workshops will thus continue in the future. In this way, the educational initiative "Haus der kleinen Forscher" seeks to make an effective contribution to professionalising early childhood educators and primary school teachers.

# 4 Outlook – Organisational Development in Educational Institutions

A longitudinal survey conducted by the Foundation among educators across all subject areas shows that the impact of the professional development workshops correlates with organisational characteristics (Stiftung Haus der kleinen Forscher, 2020b, 2022). The pilot evaluations mentioned above likewise show how much the transfer of training content to pedagogical practice depends on factors deriving from the educational institution as a whole, in particular the structural and cultural framework, the leader and team cooperation. This is why the Foundation has focused on the issue of organisational development in early childhood education and care centres for several years now. As early as 2016, the Foundation's Education for Sustainable Development programme included activities aimed at supporting the entire institution in the sense of a "whole institution approach" (Ferreira, Ryan & Tilbury, 2006), while special programmes were also developed for early childhood education and care centre leaders in this connection (Stiftung Haus der kleinen Forscher, 2019b). Since 2019, the Foundation has boosted its contribution to the development of pre-primary STEM education for sustainable development by initiating two additional projects.

The model programme "KiQ" (project duration: 03/2019–12/2022, funded by the Federal Ministry of Education and Research) aims to develop and test support activities for anchoring exploration- and inquiry-based learning at early childhood education and care centres. The programme has been undergoing testing since September 2020 at around 90 institutions in four selected model regions in Germany[66] involving approximately 180 early childhood education and care centre leaders and educators. Extending over several months, this professional development programme dedicated to the integration of STEM education in day-to-day routine pursues a system-oriented approach to the quality development of exploration- and inquiry-based learning which interlinks the development of competencies at the personal level with the conscious consideration of overall organisational aspects. "KiQ" thus supports the further development of the early childhood education and care centre at the organisational level.

"KiQ" programme activities seek to anchor the Foundation's pedagogical approach on a lasting basis by means of conscious planning, steering and implementation of an institution-specific change process at the participating early childhood education and care centres. Activities are to be implemented both at the level of organisational culture (e.g., values embraced in practice) and at

---

66 Baden-Württemberg, Hamburg, North Rhine-Westphalia and Saxony

the level of organisational structure (e.g., communication structures, routines and procedures). Meanwhile, the early childhood education and care centre leader expands their competence to shape the change process at the institution in such a way that this promotes the implementation of the Foundation's pedagogical approach in day-to-day work. At the same time, the educator is supported in implementing the pedagogical concept in its interaction with the children by means of ongoing qualification activities. In addition, the educator is strengthened in its role as multiplier, subsequently sharing the knowledge he/she acquires with team members. In this process, team members exercise and enhance their skills in implementing exploration- and inquiry-based learning in day-to-day interactions with the children.

The "KiQ" model programme involves extensive scientific monitoring. This consists of an internal evaluation and an external accompanying study. The results are to be published in 2023 as part of the Foundation's academic publication series.

Also in spring 2019, the "Haus der kleinen Forscher" Foundation together with the Robert Bosch Stiftung GmbH launched the cooperation project "Forum KITA-Entwicklung" with the aim of providing fresh stimuli for quality development at early childhood education and care centres. The focus of the project (03/2019-02/2023) is on how early childhood education and care centres can make use of processes of organisational development to grow as educational organisations, thereby increasing their organisational learning capacity and improving the quality of education. With its goals of "understanding, networking, changing", the project clusters specialist expertise and models, testing practical measures for early childhood education and care centre development and drawing on the results obtained from specialist research and impact measurement to develop recommendations for action, agenda-setting and political communication. The project has obtained a large number of expert reports and organises discussion of these at specialist events involving academics, practitioners and politicians as well as the representatives of associations and other organisations.

An exemplary practical support tool for early childhood education and care centre development for educators and leaders developed as part of this project is to undergo practical field testing in summer/autumn 2022. This tool is intended to be a first step in the direction of early childhood education and care centre development and promote further organisational change processes, also in the long term. The testing of the tool is being scientifically accompanied both internally at the Foundation and by external research partners. The results of the project are to be published in 2023.

With its programmes to support further development of the institution as a whole, the Foundation seeks to support educational institutions in sustainably developing as places of exploration- and inquiry-based learning, also in the field of early computer science education. The aim is to create conducive learning environments for children and instil positive attitudes towards STEM education for sustainable development, including computer science, as well as initiating a basic understanding among both children and educators. Together with their supporters – the early childhood educators and primary school teachers – the aim is for the children to take pleasure in discovering and understanding their living environment.

# References

## Introduction –

"Haus der kleinen Forscher" Foundation

Anders, Y. & Ballaschk, I. (2014). Studie zur Untersuchung der Reliabilität und Validität des Zertifizierungsverfahrens der Stiftung "Haus der kleinen Forscher". In Stiftung Haus der kleinen Forscher (Ed.), *Wissenschaftliche Untersuchungen zur Arbeit der Stiftung "Haus der kleinen Forscher"* (Vol. 6, pp. 35–116). Schaffhausen: Schubi Lernmedien AG. Available at haus-der-kleinen-forscher.de/de/wissenschaftliche-begleitung

Deutsche Telekom Stiftung & Stiftung Haus der kleinen Forscher. (2021). *Lasst den Forschergeist frei! – Ausgezeichnete Projekte des Kita-Wettbewerbs Forschergeist 2020*. Bonn/Berlin: Deutsche Telekom Stiftung/Stiftung Haus der kleinen Forscher. Available at forschergeist-wettbewerb.de

Deutsches Jugendinstitut & Weiterbildungsinitiative Frühpädagogische Fachkräfte (Eds.). (2014). *Leitung von Kindertageseinrichtungen. Grundlagen für die kompetenzorientierte Weiterbildung* (WiFF Wegweiser Weiterbildung, Vol. 10). München: WiFF.

DIVSI (2015). *DIVSI U9-Studie: Kinder in der digitalen Welt*. Retrieved March, 2022, from https://www.divsi.de/publikationen/studien/divsi-u9-studie-kinder-der-digitalen-welt/

Eickelmann, B. (2015). *Bildungsgerechtigkeit 4.0 – ICILS 2013: Grundlage für eine neue Debatte zur Bildungsgerechtigkeit*. Retrieved March, 2022, from https://www.boell.de/de/2015/04/27/bildungsgerechtigkeit

"Haus der kleinen Forscher" Foundation. (2018). *Early Science Education – Goals and Process-Related Quality Criteria for Science Teaching* (Scientific Studies on the Work of the "Haus der kleinen Forscher" Foundation, Vol. 5). Opladen, Berlin, Toronto: Verlag Barbara Budrich. Available at haus-der-kleinen-forscher.de/de/wissenschaftliche-begleitung

Rank, A., Wildemann, A., Pauen, S., Hartinger, A., Tietze, S. & Kästner, R. (2018). Early Steps into Science and Literacy – EASI Science-L. Naturwissenschaftliche Bildung in der Kita: Gestaltung von Lehr-Lern-Situationen, sprachliche Anregungsqualität und sprachliche sowie naturwissenschaftliche Fähigkeiten der Kinder. In Stiftung Haus der kleinen Forscher (Ed.), *Wirkungen naturwissenschaftlicher Bildungsangebote auf pädagogische Fachkräfte und Kinder* (Wissenschaftliche Untersuchungen zur Arbeit der Stiftung "Haus der kleinen Forscher", Vol. 10, pp. 138–251). Opladen, Berlin, Toronto: Verlag Barbara Budrich. Available at haus-der-kleinen-forscher.de/de/wissenschaftliche-begleitung

Skorsetz, N., Öz, L., Schmidt, J. K. & Kucharz, D. (2020). Entwicklungsverläufe von pädagogischen Fach- und Lehrkräften in der frühen MINT-Bildung. In Stiftung Haus der kleinen Forscher (Ed.), *Professionalisierung pädagogischer Fach- und Lehrkräfte in der frühen MINT-Bildung* (Wissenschaftliche Untersuchungen zur Arbeit der Stiftung "Haus der kleinen Forscher", Vol. 13, pp. 46–125). Opladen, Berlin, Toronto: Barbara Budrich. Available at haus-der-kleinen-forscher.de/de/wissenschaftliche-begleitung

Steffensky, M., Anders, Y., Barenthien, J., Hardy, I., Leuchter, M., Oppermann, E. et al. (2018). Early Steps into Science – EASI Science. Wirkungen früher naturwissenschaftlicher Bildungsangebote auf die naturwissenschaftlichen Kompetenzen von Fachkräften und Kindern. In Stiftung Haus der kleinen Forscher (Ed.), *Wirkungen naturwissenschaftli-*

cher Bildungsangebote auf pädagogische Fachkräfte und Kinder (Wissenschaftliche Untersuchungen zur Arbeit der Stiftung "Haus der kleinen Forscher", Vol. 10, pp. 50–136). Opladen, Berlin, Toronto: Verlag Barbara Budrich. Available at haus-der-kleinen-forscher.de/de/wissenschaftliche-begleitung

Stiftung Haus der kleinen Forscher (Ed.). (2015). *Wissenschaftliche Untersuchungen zur Arbeit der Stiftung "Haus der kleinen Forscher"* (Vol. 7). Schaffhausen: SCHUBI Lernmedien AG. Available at haus-der-kleinen-forscher.de/de/wissenschaftliche-begleitung

Stiftung Haus der kleinen Forscher (Ed.). (2017a). *Frühe mathematische Bildung – Ziele und Gelingensbedingungen für den Elementar- und Primarbereich* (Wissenschaftliche Untersuchungen zur Arbeit der Stiftung "Haus der kleinen Forscher", Vol. 8). Opladen, Berlin, Toronto: Verlag Barbara Budrich. Available at haus-der-kleinen-forscher.de/de/wissenschaftliche-begleitung

Stiftung Haus der kleinen Forscher. (2017b). *Monitoring-Bericht 2016/2017 der Stiftung "Haus der kleinen Forscher"*. Berlin: Stiftung Haus der kleinen Forscher. Available at haus-der-kleinen-forscher.de/de/wissenschaftliche-begleitung

Stiftung Haus der kleinen Forscher (Ed.). (2018a). *Frühe informatische Bildung – Ziele und Gelingensbedingungen für den Elementar- und Primarbereich* (Wissenschaftliche Untersuchungen zur Arbeit der Stiftung "Haus der kleinen Forscher", Vol. 9). Opladen, Berlin, Toronto: Verlag Barbara Budrich. Available at haus-der-kleinen-forscher.de/de/wissenschaftliche-begleitung

Stiftung Haus der kleinen Forscher (Ed.). (2018b). *Wirkungen naturwissenschaftlicher Bildungsangebote auf pädagogische Fachkräfte und Kinder* (Wissenschaftliche Untersuchungen zur Arbeit der Stiftung "Haus der kleinen Forscher", Vol. 10). Opladen, Berlin, Toronto: Verlag Barbara Budrich. Available at haus-der-kleinen-forscher.de/de/wissenschaftliche-begleitung

Stiftung Haus der kleinen Forscher. (2019a). *Pädagogischer Ansatz der Stiftung "Haus der kleinen Forscher"* (6th, completely revised edition). Berlin: Stiftung Haus der kleinen Forscher. Available at haus-der-kleinen-forscher.de/de/fortbildungen

Stiftung Haus der kleinen Forscher (Ed.). (2019b). *Zieldimensionen für Multiplikatorinnen und Multiplikatoren früher MINT-Bildung* (Wissenschaftliche Untersuchungen zur Arbeit der Stiftung "Haus der kleinen Forscher", Vol. 11). Opladen, Berlin, Toronto: Verlag Barbara Budrich. Available at haus-der-kleinen-forscher.de/de/wissenschaftliche-begleitung

Stiftung Haus der kleinen Forscher (Ed.). (2019c). *Frühe Bildung für nachhaltige Entwicklung – Ziele und Gelingensbedingungen* (Wissenschaftliche Untersuchungen zur Arbeit der Stiftung "Haus der kleinen Forscher", Vol. 12). Opladen, Berlin, Toronto: Barbara Budrich. Available at haus-der-kleinen-forscher.de/de/wissenschaftliche-begleitung

Stiftung Haus der kleinen Forscher. (2020a). *Monitoring-Bericht 2018/2019 der Stiftung "Haus der kleinen Forscher"*. Berlin: Stiftung Haus der kleinen Forscher. Available at haus-der-kleinen-forscher.de/de/wissenschaftliche-begleitung

Stiftung Haus der kleinen Forscher. (2020b). *Zertifizierung für Kitas, Horte und Grundschulen. So wird Ihre Einrichtung ein "Haus der kleinen Forscher"*. Berlin: Stiftung Haus der kleinen Forscher. Available at haus-der-kleinen-forscher.de/de/zertifizierung

Stiftung Haus der kleinen Forscher. (2022). *Monitoring-Bericht 2020/2021 der Stiftung "Haus der kleinen Forscher"*. Berlin: Stiftung Haus der kleinen Forscher. Available at haus-der-kleinen-forscher.de/de/wissenschaftliche-begleitung

Strehmel, P. & Ulber, D. (Eds.). (2017). *Kitas leiten und entwickeln. Ein Lehrbuch zum Kita-Management*. Stuttgart: Kohlhammer.

# Goal Dimensions of Computer Science Education at the Elementary and Primary Level –

Nadine Bergner, Hilde Köster, Johannes Magenheim, Kathrin Müller, Ralf Romeike, Ulrik Schroeder, Carsten Schulte

Adams, J. C. & Webster, A. R. (2012). What do students learn about programming from game, music video, and storytelling projects? *Proceedings of the 43rd ACM Technical Symposium on Computer Science Education* (pp. 643–648). ACM Press. doi:10.1145/2157136.2157319

Albers, C., Magenheim, J. & Meister, D. (Eds.). (2011). *Schule in der digitalen Welt: medienpädagogische Ansätze und Schulforschungsperspektiven* (Medienbildung und Gesellschaft). Wiesbaden: VS Verlag für Sozialwissenschaften.

Ambler, S. W. (1998). *Building object applications that work: your step-by-step handbook for developing robust systems with object technology* (No. 9). Cambridge, New York: Cambridge University Press.

Anders, Y., Hardy, I., Pauen, S. & Steffensky, M. (2017). Goals of Science Education Between the Ages of Three and Six and Their Assessment. In "Haus der kleinen Forscher" Foundation (Ed.), *Early Science Education – Goals and Process-Related Quality Criteria for Science Teaching* (Scientific Studies on the Work of the "Haus der kleinen Forscher" Foundation, Vol. 5, pp. 30–99). Opladen, Berlin, Toronto: Verlag Barbara Budrich. Available at haus-der-kleinen-forscher.de/de/wissenschaftliche-begleitung

Anders, Y., Hardy, I., Sodian, B. & Steffensky, M. (2017). Goals of Science Education at Primary School Age and Their Assessment. In "Haus der kleinen Forscher" Foundation (Ed.), *Early Science Education – Goals and Process-Related Quality Criteria for Science Teaching* (Scientific Studies on the Work of the "Haus der kleinen Forscher" Foundation, Vol. 5, pp. 100–171). Opladen, Berlin, Toronto: Verlag Barbara Budrich. Available at haus-der-kleinen-forscher.de/de/wissenschaftliche-begleitung

Bagozzi, R. P. (2007). The legacy of the technology accepture model and a proposal for a paradigm shift. *Journal of the Association for Information Systems 8, 4*, 244–254.

Bell, T. (2000). A low-cost high-impact computer science show for family audiences. In *Proceedings 23rd Australasian Computer Science Conference. ACSC 2000 (Cat. No. PR00518*, pp. 10-16). IEEE.

Bell, T., Rosamond, F. & Casey, N. (2012). Computer Science Unplugged and Related Projects in Math and Computer Science Popularization (Lecture Notes in Computer Science). In H.L. Bodlaender, R. Downey, F.V. Fomin & D. Marx (Eds.), *The Multivariate Algorithmic Revolution and Beyond* (pp. 398–456). Berlin, Heidelberg: Springer. doi:10.1007/978-3-642-30891-8_18

Bell, T., Witten, I. H. & Fellows, M. (1998). *Computer Science Unplugged: Off-line activities and games for all ages*. Computer Science Unplugged.

Bender, E., Hubwieser, P., Schaper, N., Margaritis, M., Berges, M., Ohrndorf, L. et al. (2015). Towards a Competency Model for Teaching Computer Science. *Peabody Journal of Education, 90* (4), 519–532. doi:10.1080/0161956X.2015.1068082

Bender, E., Schaper, N., Caspersen, M. E., Margaritis, M. & Hubwieser, P. (2016). Identifying and formulating teachers' beliefs and motivational orientations for computer science teacher education. *Studies in Higher Education, 41* (11), 1–16. doi:10.1080/03075079.2015.1004233

Benz, C., Grüßing, M., Lorenz, J. H., Selter, C. & Wollring, B. (2017). Zieldimensionen mathematischer Bildung im Elementar- und Primarbereich. In Stiftung Haus der kleinen Forscher (Eds.), *Frühe mathematische Bildung – Ziele und Gelingensbedingungen für den Elementar- und Primarbereich* (Wissenschaftliche Untersuchungen zur Arbeit der Stiftung "Haus der kleinen Forscher", Vol. 8, pp. 32–177). Opladen, Berlin, Toronto: Verlag Barbara Budrich. Available at haus-der-kleinen-forscher.de/de/wissenschaftliche-begleitung

Bergner, N. (2015). *Konzeption eines Informatik-Schülerlabors und Erforschung dessen Effekte auf das Bild der Informatik bei Kindern und Jugendlichen.* Aachen: Lehr- und Forschungsgebiet Informatik 9, RWTH Aachen University.

Beutner, M., Kundisch, D., Magenheim, J. & Zoyke, A. (2014). Support, Supervision, Feedback and Lectures Role in the use of the Classroom Response Systems PINGO. *Proceedings of World Conference on E-Learning in Corporate, Government, Healthcare, and Higher Education 2014* (pp. 1210–1217).

Biggs, J. B. & Collis, K. F. (1982). *Evaluating the quality of learning: the SOLO taxonomy (Structure of the Observed Learning Outcome).* New York: Academic Press.

Blömeke, S. (Ed.). (2008). *Professionelle Kompetenz angehender Lehrerinnen und Lehrer: Wissen, Überzeugungen und Lerngelegenheiten deutscher Mathematikstudierender und -referendare ; erste Ergebnisse zur Wirksamkeit der Lehrerausbildung.* Münster: Waxmann.

Blömeke, S., Kaiser, G. & Lehmann, R. (2011). Messung professioneller Kompetenz angehender Lehrkräfte: "Mathematics Teaching in the 21st Century" und die IEAStudie TEDS-M. (Fachdidaktische Forschungen). In H. Bayrhuber, U. Harms, B. Muszynski, B. Ralle, M. Rothgangel, L.-H. Schön et al. (Eds.), *Empirische Fundierung in den Fachdidaktiken* (pp. 9–26). Münster: Waxmann.

Blumer, H. (1969). *Symbolic Interactionism: Perspective and Method.* Berkeley, Los Angeles: University of California Press.

Boles, D. (2005). *Spielerisches Erlernen der Programmierung mit dem Java-Hamster-Modell. Capturing the interest of the uninterested* (pp. 243–252). Presented at Lecture Notes in Informatics (LNI)-Proceedings 60, Wellington, New Zealand.

Borowski, C. & Diethelm, I. (2009). Kinder auf dem Wege zur Informatik: Programmieren in der Grundschule. *13. GI-Fachtagung-Informatik und Schule.* Presented at INFOS, Berlin: LOG IN.

Borowski, C., Diethelm, I. & Mesaros, A.-M. (2010). *Informatische Bildung im Sachunterricht in der Grundschule.* Available at www.widerstreit-sachunterricht.de

Borowski, C., Diethelm, I., & Wilken, H. (2016). What children ask about computers, the Internet, robots, mobiles, games etc. In *Proceedings of the 11th Workshop in Primary and Secondary Computing Education* (pp. 72-75).

Börstler, J., & Schulte, C. (2005). Teaching object oriented modelling with CRC cards and roleplaying games. In *Proceedings WCCE* (Vol. 5). Available at http://citeseerx.ist.psu.edu/viewdoc/download?doi=10.1.1.437.3492&rep=rep1&type=pdf

Brabrand, C. & Dahl, B. (2009). Using the SOLO taxonomy to analyze competence progression of university science curricula. *Higher Education, 58* (4), 531–549. doi:10.1007/s10734-009-9210-4

Brandhofer, G. (2014). Ein Gegenstand "Digitale Medienbildung und Informatik" – notwendige Bedingung für digitale Kompetenz? *R&E-SOURCE (1).*

Brandt-Pook, H. & Kollmeier, R. (2008). *Softwareentwicklung kompakt und verständlich Wie Softwaresysteme entstehen.* Wiesbaden: Vieweg +Teubner/GWV Fachverlage GmbH, Wiesbaden.

Brauner, P. (2009). *Konzeption, Entwicklung und Analyse eines greifbaren Turtles in Hinblick auf die Steigerung der Computerselbstwirksamkeit von Schülerinnen und Schülern.* Diplomarbeit. Aachen: RWTH Aachen.

Breier, N. (2004a). *Informatik und die klassischen Naturwissenschaften – Partner oder Kontrahenten?* Available at https://www.ew.uni-hamburg.de/ueber-die-fakultaet/personen/breier/files/mnu-pdf.pdf

Breier, N. (2004b). Stand und Perspektive der informatischen Bildung. *1. Fachtagung der GI-Fachgruppe Hamburger Informatiklehrerinnen und -lehrer.*

Breiter, A., Welling, S. & Stolpmann, B. E. (2010). *Medienkompetenz in der Schule: Integration von Medien in den weiterführenden Schulen in Nordrhein-Westfalen* (Schriftenreihe Medienforschung der LfM). Berlin: Vistas Verlag.

Brennan, K. (2013). Learning Computing through Creating and Connecting. *Computer, 46* (9), 52–59. doi:10.1109/MC.2013.229

Brinda, T., Diethelm, I., Gemulla, R., Romeike, R., Schöning, J. & Schulte, C. (2016). *Dagstuhl-Erklärung: Bildung in der digitalen vernetzten Welt.* GI – Gesellschaft für Informatik e.V. Retrieved March, 2022, from https://gi.de/themen/beitrag/dagstuhl-erklaerung-bildung-in-der-digital-vernetzten-welt

Bröker, K., Kastens, U. & Magenheim, J. (2014). Competences of Undergraduate Computer Science Students. In T. Brinda, N. Reynolds & R. Romeike (Eds.), *Key Competencies in Informatics and ICT* (pp. 77–96). Presented at KEYCIT 2014, Potsdam.

Brown, N. C. C., Sentance, S., Crick, T. & Humphreys, S. (2014). Restart: The Resurgence of Computer Science in UK Schools. *ACM Transactions on Computing Education, 14* (2), 1–22. doi:10.1145/2602484

Buchholz, M., Saeli, M. & Schulte, C. (2013). PCK and reflection in computer science teacher education. *Proceedings of the 8th Workshop in Primary and Secondary Computing Education* (pp. 8–16). ACM Press. doi:10.1145/2532748.2532752

Bussmann, H. & Heymann, H. W. (1987). Computer und Allgemeinbildung. *Vierteljahreszeitschrift für Erziehung und Gesellschaft, 27* (1), 2–39.

Carlsen, W. (2002). Domains of Teacher Knowledge. In J. Gess-Newsome & N.G. Lederman (Eds.), *Examining Pedagogical Content Knowledge* (Vol. 6, pp. 133–144). Dordrecht: Kluwer Academic Publishers. Retrieved March, 2022, from http://link.springer.com/10.1007/0-306-47217-1_5

Caspersen, M. E. & Kolling, M. (2009). STREAM: A First Programming Process. *ACM Transactions on Computing Education, 9* (1), 1–29. doi:10.1145/1513593.1513597

Chaudron, S. (2015). *Young Children (0-8) and digital technology: a qualitative exploratory study across seven countries.* (European Commission, Joint Research Centre & Institute for the Protection and the Security of the Citizen, Ed.). Luxembourg: Publications Office. Retrieved March, 2022, from http://dx.publications.europa.eu/10.2788/00749

Claus, V. (1977). Informatik an der Schule: Begründungen und allgemeinbildender Kern (Schriftenreihe des Instituts für Didaktik der Mathematik der Universität Bielefeld). In H. Bauersfeld, M. Otte & H.G. Steiner (Eds.), *Informatik im Unterricht der Sekundarstufe II: Grundfragen, Probleme und Tendenzen mit Bezug auf allgemeinbildende und berufsqualifizierte Ausbildungsgänge* (Vol. 1, pp. 19–33). Bielefeld.

Claus, V. & Schwill, A. (Eds.). (2006). *Duden Informatik A – Z: Fachlexikon für Studium, Ausbildung und Beruf* (Duden) (4th ed.). Mannheim: Dudenverlag.

Computing at School Working Group. (2013). *Computing in the national curriculum: A guide for primary teachers*. Available at www.computingatschool.org.uk/primary

Crutzen, C. K. M. (2000). *Interaction, a world of differences. A vision on informatics from the perspective of gender studies*. Open Universiteit Nederland.

CSTA K-12 Computer Science Standards. (2011). Available at https://www.csteachers.org/page/about-csta-s-k-12-nbsp-standards

Davis, E. (2008). Teaching Elementary Teachers' Ideas about Effective Science Teaching: A Longitudinal Study. *Proceedings of the 8th international conference for the learning sciences* (Vol. 1, pp. 199–206).

Department for Education. (2013). *Statutory guidance – National curriculum in England: computing programmers of study*.

Dewey, J. (1938). *Experience and education*. New York: Simon & Schuster.

Deutschschweizer Erziehungsdirektion. (2015). *Lehrplan 21 zur Einführung in den Kantonen*.

Deutschschweizer Erziehungsdirektoren-Konferenz (Ed.). (2016). *Lehrplan 21 – Medien und Informatik*. Luzern.

Dickins, R., Nielsen, S., Barden, E. & Lamont, H. (2015). *Lift-the-flap computers and coding*. Usborne Publishing.

DIVSI. (2015). *DIVSI U9-Studie: Kinder in der digitalen Welt*. Retrieved March, 2022, from https://www.divsi.de/publikationen/studien/divsi-u9-studie-kinder-der-digitalen-welt/

Döbeli Honegger, B. (2010). ICT im Hosensack – Informatik im Kopf? Presented at *25 Jahre Schulinformatik: Zukunft mit Herkunft,* Österreichische Computer-Gesellschaft.

Döbeli Honegger, B. (2016). *Mehr als 0 und 1: Schule in einer digitalisierten Welt*. hep der bildungsverlag.

Dohmen, M., Magenheim, J. & Engbring, D. (2009). Kreativer Einstieg in die Programmierung – Alice im Informatik Anfangsunterricht. *Informatische Bildung in Theorie und Praxis* (Vol. 13, pp. 69–80). Presented at INFOS, LOG IN.

Döhrmann, M., Kaiser, G. & Blömeke, S. (2010). Messung des mathematischen und mathematikdidaktischen Wissens: Theoretischer Rahmen und Teststruktur. In S. Blömeke, G. Kaiser & R. Lehmann (Eds.), *TEDS-M 2008 – Professionelle Kompetenz und Lerngelegenheiten angehender Primarstufenlehrkräfte im internationalen Vergleich*. Münster: Waxmann.

Döhrmann, M., Kaiser, G. & Blömeke, S. (2012). The conceptualisation of mathematics competencies in the international teacher education study TEDS-M. *ZDM, 44* (3), 325–340. doi:10.1007/s11858-012-0432-z

Duncan, C. & Bell, T. (2015). A Pilot Computer Science and Programming Course for Primary School Students. In *Proceedings of the Workshop in Primary and Secondary Computing Education* (pp. 39–48). ACM Press. doi:10.1145/2818314.2818328

Dworschak, M. (2015). Wie man Lehrer fernsteuert. *Der Spiegel, 44/2015*, 114–116.

Eberle, F. (1996). *Didaktik der Informatik bzw. einer informations- und kommunikationstechnologischen Bildung auf der Sekundarstufe II*. Aarau: Verlag für Berufsbildung Sauerländer.

Eickel, J., Brauer, W., Claus, V., Deussen, P., Haake, W., Hosseus, W. et al. (1969). Zielsetzungen und Inhalte des Informatikunterrichts. In Im Auftrag des Fachausschusses "Ausbildung" des Gesellschaft für Informatik (Ed.), *Zentralblatt für Didaktik der Mathematik* (Vol. 1, pp. 35–43). Klett.

Engbring, D. & Selke, H. (2013). Informatik und Gesellschaft als Gebiet der Informatik. Presented at *HDI 2012–Informatik für eine nachhaltige Zukunft: 5. Fachtagung Hochschuldidaktik der Informatik*; November, 6th–7th, 2012, Universität Hamburg (Vol. 5, pp. 111–116).

Eulenberger, J. (2015). Die Persönlichkeitsmerkmale von Personen im Kontext des Lehrer_innenberufs. Presented at *DIW-SOEP – German Socio Economic Panel Study* (SOEP).

Fernandez, C. (2014). Knowledge Base for Teaching and Pedagogical Content Knowledge (PCK): Some useful Models and Implications for Teachers' Training. *Training Problems of education in the 21st century, 60,* 79–100.

Fessakis, G. & Karakiza, T. (2011). Pedagogical beliefs and attitudes of computer science teachers in Greece. *Themes in Science & Technology Education, 4*(2), 75–88.

Feurzeig, W., Papert, S. A. & Lawler, B. (1970). Programming-languages as a conceptual framework for teaching mathematics. *Interactive Learning Environments, 19*(5), 487–501. doi:10.1080/10494820903520040

Fischer, G., Giaccardi, E., Ye, Y., Sutcliffe, A. G. & Mehandjiev, N. (2004). Meta-design: a manifesto for end-user development. *Communications of the ACM, 47*(9), 33–38. doi:10.1145/1015864.1015884

Flannery, L. P., Silverman, B., Kazakoff, E. R., Bers, M. U., Bontá, P. & Resnick, M. (2013). *Designing ScratchJr: support for early childhood learning through computer programming* (pp. 1–10). ACM Press. doi:10.1145/2485760.2485785

Forneck, H.-J. (1992). *Bildung im informationstechnischen Zeitalter: Untersuchung der fachdidaktischen Entwicklung der informationstechnischen Bildung*. Aarau: Sauerländer.

Frank, H. & Meyer, I. (1974). *Kybernetische Pädagogik Schriften 1958-1972. Rechnerkunde: Elemente d. digitalen Nachrichtenverarbeitung u. ihrer Fachdidaktik* (Vol. 5). Stuttgart, Berlin, Köln, Mainz: Kohlhammer.

Fröbel, F. (1826). The Education of Man. (Reprint: 1974). A. M. Kelley. https://books.google.de/books/about/The_Education_of_Man.html?id=SLJEAQAAMAAJ&redir_esc=y

Fuller, U., Riedesel, C., Thompson, E., Johnson, C. G., Ahoniemi, T., Cukierman, D. et al. (2007). Developing a computer science-specific learning taxonomy. *ACM SIGCSE Bulletin, 39*(4), 152-170. doi:10.1145/1345443.1345438

Fulton, K. L. (1999). *How teachers' beliefs about teaching and learning are reflected in their use of technology: case study from urban middle schools*. Unpublished doctoral dissertation. University of Maryland.

Gallenbacher, J. (2009). Abenteuer Informatik – "Abenteuer begreifen" wörtlich gemeint. *INFOS*.

Gander, W., Petit, A., Berry, G., Demo, B., Vahrenhold, J., McGettrick, A. et al. (2013). Informatics education: Europe cannot afford to miss the boat. *Report of the joint Informatics Europe & ACM Europe Working Group on Informatics Education*.

Gesellschaft für Didaktik des Sachunterrichts (Ed.). (2013). *Perspektivrahmen Sachunterricht*. Bad Heilbrunn: Verlag Julius Klinkhardt.

GI – Gesellschaft für Informatik e.V. (2008). *Bildungsstandards Informatik für die Sekundarstufe I: Empfehlungen der Gesellschaft für Informatik e.V.* (Vol. 28). LOG IN.

GI – Gesellschaft für Informatik e.V. (2016a). *Bildungsstandards Informatik für die Sekundarstufe II, Empfehlungen der Gesellschaft für Informatik e.V.* (Arbeitskreis "Bildungsstandards SII"). LOG IN.

GI – Gesellschaft für Informatik e.V. (2016b). *Empfehlungen für Bachelor- und Masterprogramme im Studienfach Informatik an Hochschulen*. Bonn.

GI – Gesellschaft für Informatik e.V. (2019). *Kompetenzen für informatische Bildung im Primarbereich, Empfehlungen der Gesellschaft für Informatik e.V.* (Arbeitskreis "Bildungsstandards Informatik im Primarbereich"). LOG IN (39).

Gibson, J. P. (2012). Teaching graph algorithms to children of all ages. In T. Lapidot, J. Gal-Ezer, M.E. Caspersen & O. Hazzan (Eds.), *Proceedings of the 17th ACM annual conference on Innovation and technology in computer science education* (pp. 34–39). New York: ACM Press. Retrieved March, 2022, from http://dl.acm.org/citation.cfm?doid=2325296.2325308

Gil, J., Schwarz, B. B. & Asterhan, C. S. C. (2007). Intuitive moderation styles and beliefs of teachers in CSCL-based argumentation. *Association for Computational Linguistics*, 222–231. doi:10.3115/1599600.1599643

Grafe, S. & Breiter, A. (2014). Modeling and Measuring Pedagogical Media Competencies of Pre-Service Teachers (M3K). In C. Kuhn, M. Toepper & O. Zlatkin-Troitschanskaia (Eds.), *Current International State and Future Perspectives on Competences Assessment in Higher Education* (pp. 76–80). Presented at Report from the KoKoHs Affiliated Group Meeting at the AERA Conference, Philadelphia (USA).

Gujberova, M. & Kalas, I. (2013). Designing productive gradations of tasks in primary programming education. In *Proceedings of the 8th workshop in primary and secondary computing education* (pp. 108–117).

Gutnick, A. L., Robb, M., Takeuchi, L., Kotler, J., Bernstein, L. & Levine, M. H. (2011). *Always connected: The new digital media habits of young children*. Retrieved March, 2022, from https://joanganzcooneycenter.org/wp-content/uploads/2011/03/jgcc_alwaysconnected.pdf

Hartig, J. & Klieme, E. (2006). Kompetenz und Kompetenzdiagnostik. In K. Schweizer (Ed.), *Leistung und Leistungsdiagnostik* (pp. 127–143). Berlin/Heidelberg: Springer-Verlag. https://doi.org/10.1007/3-540-33020-8_9

Hartmann, S. & Schecker, H. (2005). Bietet Robotik Mädchen einen Zugang zu Informatik, Technik und Naturwissenschaft? – Evaluationsergebnisse zu dem Projekt "Roberta". *Zeitschrift für Didaktik der Naturwissenschaften, 11*, 7–19.

Hattie, J. (2009). *Visible learning: a synthesis of over 800 meta-analyses relating to achievement*. London, New York: Routledge.

"Haus der kleinen Forscher" Foundation (Ed.). (2018). *Early Science Education – Goals and Process-Related Quality Criteria for Science Teaching* (Scientific Studies on the Work of the "Haus der kleinen Forscher" Foundation, Vol. 5). Opladen, Berlin, Toronto: Verlag Barbara Budrich. Available at haus-der-kleinen-forscher.de/de/wissenschaftliche-begleitung

Hauser, B. & Rechsteiner, K. (2011). Frühe Mathematik: Geführtes Spiel oder Training? *4 bis 8–Schweizerische Fachzeitschrift für Kindergarten und Unterstufe, 5*, 28–30.

Heimann, P., Otto, G. & Schulz, W. (1979). *Unterricht: Analyse und Planung* (Auswahl Reihe B) (10th ed.). Hannover: Schroedel.

Herczeg, M. (2009). *Software-Ergonomie: Theorien, Modelle und Kriterien für gebrauchstaugliche interaktive Computersysteme* (Lehrbuchreihe interaktive Medien) (3rd ed.). München: Oldenbourg.

Hettlage, R. & Steinlin, M. (2006). The Critical Incident Technique in Knowledge Management Related Contexts. *Ingenious Peoples Knowledge*, 1-18.

Heymann, H.-W. (1997). Allgemeinbildung als Aufgabe der Schule und als Maßstab für Fachunterricht. *Allgemeinbildung und Fachunterricht* (pp. 7–17). Hamburg: Bergmann + Helbig.

Hoppe, H. U. & Löthe, H. (1984). *Problemlösen und Programmieren mit LOGO Ausgewählte Beispiele aus Mathematik und Informatik*. Wiesbaden: Vieweg+Teubner Verlag.

Hubwieser, P. (2007). *Didaktik der Informatik: Grundlagen, Konzepte, Beispiele*. Berlin, Heidelberg: Springer.

Hubwieser, P. & Broy, M. (1997). Ein neuer Ansatz für den Informatikunterricht am Gymnasium. *LOG IN, (3/4)*, 42–47.

Hubwieser, P., Magenheim, J., Mühling, A. & Ruf, A. (2013). Towards a conceptualization of pedagogical content knowledge for computer science. In *Proceedings of the ninth annual international ACM conference on International computing education research* (pp. 1–8).

Hubwieser, P., Schubert, S., Armoni, M., Brinda, T., Dagiene, V., Diethelm, I. et al. (2011). Computer science/informatics in secondary education. In *Proceedings of the 16th annual conference reports on Innovation and technology in computer science education – working group reports* (pp. 19–38).

Humbert, L. & Puhlmann, H. (2004). Essential ingredients of literacy in informatics. *Informatics and Student Assessment. Concepts of empirical research and standardisation of measurement in the area of didactics* (Vol. 1, pp. 65–76).

Initiative D21 e.V. (2008). *Bildung via Internet: Wie vernetzt sind unsere Kinder? Eine Sonderstudie im Rahmen des (N)Onliner Atlas*. Retrieved March, 2022, from http://neu.initiatived21.eu/wp-content/uploads/alt/08_NOA/FSC_Sonderstudie_72dpi.pdf.

Isayama, D., Ishiyama, M., Relator, R. & Yamazaki, K. (2016). Computer Science Education for Primary and Lower Secondary School Students: Teaching the Concept of Automata. *ACM Transactions on Computing Education, 17*(1), 1–28. doi:10.1145/2940331

ISTE – The International Society for Technology in Education. (2008). *Standards for Teachers*. Retrieved March, 2022, from http://www.iste.org/standards/iste-standards/standards-for-students.

Kafai, Y. B. & Burke, Q. (2014). *Connected code: why children need to learn programming*. Cambridge, Massachusetts: The MIT Press.

Kay, A. & Goldberg, A. (1977). Personal Dynamic Media. *Computer, 10*(3), 31–41. doi:10.1109/C-M.1977.217672

Keil, R. (2012). Der Computer als Medium – Medien als Denkzeug des Geistes. In A. Knaut, C. Kühne, R. Rehak, S. Ullrich, C. Kurz & J. Pohle (Eds.), *Per Anhalter durch die Turing-Galaxis* (pp. 147–219). Münster: Verlagshaus Monsenstein und Vannerdat.

Keil-Slawik, R. (1994). *Softwareentwicklung: Die Gestaltung des Unsichtbaren.* Kurzfassung in der Vorlesung Informatik und Gesellschaft.

Kelleher, C. (2006). *Motivating Programming: Using storytelling to make computer programming attractive to middle school girls.* Dissertation. Pittsburgh: Carnegie Mellon University.

Kelleher, C. (2015). Looking Glass. In *Proceedings of the 46th ACM Technical Symposium on Computer Science Education* (pp. 271–271). doi:10.1145/2676723.2691873

Kelleher, C. & Pausch, R. (2005). Lowering the barriers to programming: A taxonomy of programming environments and languages for novice programmers. *ACM Computing Surveys, 37*(2), 83–137. doi:10.1145/1089733.1089734

Kind, A. (2015). Computing Attitudes: Will Teaching 2nd Grade Students Computer Science Improve their Self-Efficacy and Attitude and Eliminate Gender Gaps? *Computer Science and Self-Efficacy, 8.*

Klafki, W. (1993). Allgemeinbildung heute. *Pädagogische Welt* (Vol. 47, pp. 28–33).

Kleickmann, T. (2008). *Zusammenhänge fachspezifischer Vorstellungen von Grundschullehrkräften zum Lehren und Lernen mit Fortschritten von Schülerinnen und Schülern im konzeptuellen naturwissenschaftlichen Verständnis.* Münster: Westfälische Wilhelms-Universität Münster.

Knobelsdorf, M. & Schulte, C., (2007). Das informatische Weltbild von Studierenden. In Schubert, S. (Ed.), *Didaktik der Informatik in Theorie und Praxis – INFOS 2007 – 12. GI-Fachtagung Informatik und Schule* (pp. 69-79). Bonn: Gesellschaft für Informatik e.V.

Körber, B. & Peters, I.-R. (1988). *Grundlagen einer Informatik der Didaktik.* FU Berlin, summer term.

Kosack, W., Jeretin-Kopf, M. & Wiesmüller, C. (2015). Zieldimensionen technischer Bildung im Elementar- und Primarbereich. In Stiftung Haus der kleinen Forscher (Ed.), *Wissenschaftliche Untersuchungen zur Arbeit der Stiftung "Haus der kleinen Forscher"* (Vol. 7, pp. 30–157). Schaffhausen: Schubi Lernmedien AG. Available at haus-der-kleinen-forscher.de/de/wissenschaftliche-begleitung

Köster, H. (2006). *Freies Explorieren und Experimentieren: eine Untersuchung zur selbstbestimmten Gewinnung von Erfahrungen mit physikalischen Phänomenen im Sachunterricht.* Berlin: Logos-Verlag.

Koubek, J., Schulte, C., Schulze, P. & Witten, H. (2009). Informatik im Kontext (IniK) – Ein integratives Unterrichtskonzept für den Informatikunterricht. *INFOS* (pp. 268–279).

Krathwohl, D. R. (2002). A Revision of Bloom's Taxonomy: An Overview. *Theory Into Practice, 41*(4), 212–218. doi:10.1207/s15430421tip4104_2

Kultusministerkonferenz. (2012). *Medienbildung in der Schule.*

Kultusministerkonferenz. (2015). *Ländergemeinsame inhaltliche Anforderungen für die Fachwissenschaften und Fachdidaktiken in der Lehrerbildung.*

Kunter, M. & Baumert, J. (2011). Das COACTIV-Forschungsprogramm zur Untersuchung professioneller Kompetenz von Lehrkräften – Zusammenfassung und Diskussion. In M. Kunter, J. Baumert, W. Blum, V. Klusmann, S. Krauss, & M. Neubrand (Eds.), *Professionelle Kompetenz von Lehrkräften – Ergebnisse des Forschungsprogramms COACTIV* (pp. 345–366). Münster: Waxmann.

Lambert, L. & Guiffre, H. (2009). Computer science outreach in an elementary school. *J. Comput. Small Coll, 24,* 118–124.

Leonhardt, T. (2015). *Etablierung eines begabungsfördernden Lernumfelds für Mädchen im Bereich Informatik*. RWTH Aachen University.

Levy, S. T. & Mioduser, D. (2008). Does it "want" or "was it programmed to…"? Kindergarten children's explanations of an autonomous robot's adaptive functioning. *International Journal of Technology and Design Education, 18*(4), 337–359. doi:10.1007/s10798-007-9032-6

Libow Martinez, S. & Stager, G. (2013). *Invent to learn: making, tinkering, and engineering in the classroom*. Torrance, Calif: Constructing Modern Knowledge Press.

Linck, B., Ohrndorf, L., Schubert, S., Stechert, P., Magenheim, J., Nelles, W. et al. (2013). Competence model for informatics modelling and system comprehension. *Global Engineering Education Conference (EDUCON'13)* (pp. 85–93). IEEE. doi:10.1109/EduCon.2013.6530090

Lindmeier, A. (2011). *Modeling and measuring knowledge and competencies of teachers: a threefold domain-specific structure model for mathematics*. Münster: Waxmann.

Liukas, L. (2015). *Hello Ruby: adventures in coding*. Macmillan.

Maceli, M. & Atwood, M. E. (2011). From Human Crafters to Human Factors to Human Actors and Back Again: Bridging the Design Time – Use Time Divide. In M.F. Costabile, Y. Dittrich, G. Fischer & A. Piccinno (Eds.), *End-User Development* (Vol. 6654, pp. 76–91). Berlin, Heidelberg: Springer.

Magenheim, J. (2000). Informatiksysteme und Dekonstruktion als didaktische Kategorien: Theoretische Aspekte und unterrichtspraktische Implikationen einer systemorientierten Didaktik der Informatik. *GI-Tagung "Informatik – Ausbildung und Beruf 2000"*.

Magenheim, J. (2008). Systemorientierte Didaktik der Informatik Sozio-technische Informatiksysteme als Unterrichtsgegenstand? (Proceedings). In U. Kortenkamp, H.-G. Weigand, T. Weth & Gesellschaft für Didaktik der Mathematik (Eds.), *Informatische Ideen im Mathematikunterricht: Bericht über die 23. Arbeitstagung des Arbeitskreises "Mathematikunterricht und Informatik" in der Gesellschaft für Didaktik der Mathematik e.V. vom 23. bis 25. September 2005 in Dillingen an der Donau* (pp. 17–36). Hildesheim: Franzbecker.

Magenheim, J., Nelles, W., Rhode, T., Schaper, N., Schubert, S. & Stechert, P. (2010). Competencies for informatics systems and modeling: Results of qualitative content analysis of expert interviews. *Education Engineering (EDUCON)* (pp. 513–521). IEEE. doi:10.1109/EDUCON.2010.5492535

Magenheim, J. & Schulte, C. (2006). Social, ethical and technical issues in informatics – An integrated approach. *Education and Information Technologies, 11*(3–4), 319–339. doi:10.1007/s10639-006-9012-6

Magenheim, J., Schulte, C. & Scheel, O. (2002). Informatics and Media Education. Designing a Curriculum for Media Education in Teacher Training with Regard to Basic Areas of Informatics. In P. Barker & S. Rebelsky (Eds.), *Proceedings of ED-MEDIA 2002--World Conference on Educational Multimedia, Hypermedia & Telecommunications* (pp. 1200-1205). Denver, Colorado, USA: Association for the Advancement of Computing in Education (AACE).

Magnusson, S., Krajcik, J. & Borko, H. (1999). Nature, Sources, and Development of Pedagogical Content Knowledge for Science Teaching. In J. Gess-Newsome & N.G. Lederman (Eds.), *Examining Pedagogical Content Knowledge* (Vol. 6, pp. 95–132). Dordrecht: Kluwer Academic Publishers.

Mahr, B. (2009). Die Informatik und die Logik der Modelle. *Informatik-Spektrum, 32*(3), 228–249. doi:10.1007/s00287-009-0340-y

Mayerová, K. (2012). Pilot Activities: LEGO WeDo at Primary School. In M. Moro & D. Alimisis (Eds.), *Proceedings of 3rd International Workshop Teaching Robotics* (pp. 32–39).

Mayring, P. (2010). *Qualitative Inhaltsanalyse: Grundlagen und Techniken* (Beltz Pädagogik) (11th ed.). Weinheim: Beltz.

McKenney, S. & Voogt, J. (2010). Technology and young children: How 4–7 year olds perceive their own use of computers. *Computers in Human Behavior, 26*(4), 656–664. doi:10.1016/j.chb.2010.01.002

McNerney, T. (2004). From turtles to Tangible Programming Bricks: explorations in physical language design. *Personal and Ubiquitous Computing, 8*(5). doi:10.1007/s00779-004-0295-6

Medienwissenschaft Universität Bayreuth. (2014). *Planspiel Datenschutz (2.0) – Wer weiß was über mich im Internet?* Available at: https://medienwissenschaft.uni-bayreuth.de/inik/entwuerfe/planspiel-datenschutz-2-0/

Meyer, J. H. F. & Land, R. (2005). Threshold concepts and troublesome knowledge (2): Epistemological considerations and a conceptual framework for teaching and learning. *Higher Education, 49*(3), 373–388. doi:10.1007/s10734-004-6779-5

Mishra, P. & Koehler, M. J. (2006). Technological pedagogical content knowledge: A framework for teacher knowledge. *Teachers College Record, 108*(6), 1017–1054.

MIT. (2011). *Crickets*. Available at: https://www.media.mit.edu/projects/crickets/overview/

Mittermeir, R. (2010). Informatikunterricht zur Vermittlung allgemeiner Bildungswerte. *OCG Schriftenreihe* (Vol. 271, pp. 54–73).

Modrow, E. (2010). Informatik als technisches Fach. *springerprofessional.de, LOG IN (163/164)*, 38–42.

Morreale, P. & Joiner, D. (2011). Reaching future computer scientists. *Communications of the ACM, 54*(4), 121–124. doi:10.1145/1924421.1924448

MPFS – Medienpädagogischer Forschungsverbund Südwest. (2014a). *KIM-Studie 2014. Kinder + Medien, Computer + Internet*. Stuttgart. Available at http://www.mpfs.de/fileadmin/files/Studien/KIM/2014/KIM_Studie_2014.pdf

MPFS – Medienpädagogischer Forschungsverbund Südwest. (2014b). *miniKIM 2014. Kleinkinder und Medien*. Stuttgart. Available at http://www.mpfs.de/fileadmin/files/Studien/miniKIM/2014/Studie/miniKIM_Studie_2014.pdf

MPFS – Medienpädagogischer Forschungsverbund Südwest. (2015). *JIM-Studie 2015 – Jugend, Information, (Multi-) Media*. Stuttgart. Available at http://www.mpfs.de/fileadmin/files/Studien/JIM/2015/JIM_Studie_2015.pdf

Müller, K. (2015). What do we expect from graduates in CS? First results of a survey at university and company as part of a methodology for developing a competence model. Presented at *IFIP TC3 Working Conference "A New Culture of Learning: Computing and next Generations"* (pp. 192–201), Vilnius, Lithuania.

Myers, B. A. (1990). Taxonomies of visual programming and program visualization. *Journal of Visual Languages & Computing, 1*(1), 97–123. doi:10.1016/S1045-926X(05)80036-9

National Research Council (U.S.) & Committee for the Workshops on Computational Thinking. (2010). *Report of a workshop on the scope and nature of computational thinking*. Washington, D.C.: National Academies Press.

National Research Council (U.S.) & Committee for the Workshops on Computational Thinking. (2011). *Report of a workshop of pedagogical aspects of computational thinking*. Washington, D.C: National Academies Press.

Naur, P. (1985). Programming as theory building. *Microprocessing and microprogramming*, 15(5), 253–261.

NCTM – National Council of Teachers of Mathematics. (1989). *Curriculum and evaluation standards for school mathematics*.

NCTM – National Council of Teachers of Mathematics. (1991). *Professional standards for teaching mathematics*.

NCTM – National Council of Teachers of Mathematics. (1995). *Assessment standards for school mathematics*.

NCTM – National Council of Teachers of Mathematics. (2000). *Principles and Standards for School Mathematics*.

Neugebauer, J., Magenheim, J., Ohrndorf, L., Schaper, N. & Schubert, S. (2015). Defining Proficiency Levels of High School Students in Computer Science by an Empirical Task Analysis Results of the MoKoM Project. In A. Brodnik & J. Vahrenhold (Eds.), *Informatics in Schools. Curricula, Competences, and Competitions* (Vol. 9378, pp. 45–56). Cham: Springer International Publishing.

Ni, L. (2011). *Building professional identity as computer science teachers: supporting secondary computer science teachers through reflection and community building* (Doctoral dissertation). Georgia Institute of Technology, Atlanta, GA.

Ni, L. & Guzdial, M. (2015). Prepare and Support Computer Science (CS) Teachers: Understanding CS Teachers' Professional Identity. In *American Educational Research Association (AERA) Annual Meeting*.

OECD. (2005). *The Definition and Selection of Key Competencies. Executive Summary*.

Palaiologou, I. (2016). Children under five and digital technologies: implications for early years pedagogy. *European Early Childhood Education Research Journal*, 24(1), 5–24. doi:10.1080/1350293X.2014.929876

Papert, S. (1980). *Mindstorms: children, computers, and powerful ideas*. New York: Basic Books.

Papert, S. (1982). *Mindstorms: Kinder, Computer und Neues Lernen*. Basel: Birkhäuser Verlag.

Papert, S. (1987). *Microworlds: transforming education. Artificial intelligence and education* (Vol. 1, pp. 79–94).

Papert, S. (1993). *The children's machine: Rethinking school in the age of the computer*. New York: Basic Books.

Papert, S. & Harel, I. (1991). Situating Constructionism. *Constructionism*, 36(2), 1–11.

Papert, S. & Stark-Städele, J. (1998). *Die vernetzte Familie: Kinder und Computer*. Stuttgart: Kreuz.

Pauen, S. & Pahnke, J. (2008). Mathematische Kompetenzen im Kindergarten. Evaluation der Effekte einer Kurzzeitintervention. *Empirische Pädagogik*, 22, 193–208.

Perlman, R. (1976). *Using computer technology to provide a creative learning environment for preschool children*. MIT Cambridge: Logo Memo.

Petersen, U., Theidig, G., Bördig, J., Leimbach, T. & Flintrop, B. (2007). *Abschlussbericht Roberta*.

Plowman, L., Stevenson, O., Stephen, C. & McPake, J. (2012). Preschool children's learning with technology at home. *Computers & Education, 59*(1), 30–37. doi:10.1016/j.compedu.2011.11.014

Pohl, W., Kranzdorf, K. & Hein, H.-W. (2007). Einstieg Informatik – Aktivitäten und Erfahrungen. *Didaktik der Informatik in Theorie und Praxis–INFOS 2007–12. GI-Fachtagung Informatik und Schule.*

Portelance, D. J., Strawhacker, A. L. & Bers, M. U. (2016). Constructing the ScratchJr programming language in the early childhood classroom. *International Journal of Technology and Design Education, 26*(4), 489–504. doi:10.1007/s10798-015-9325-0

Rechenberg, P. (2010). Was ist Informatik. *25 Jahre Schulinformatik* (pp. 46–53). Österreichische Computer-Gesellschaft.

Reichert, R., Nievergelt, J. & Hartmann, W. (2005). *Programmieren mit Kara: ein spielerischer Zugang zur Informatik.* Berlin: Springer-Verlag.

Resnick, M. (2007). Sowing the Seeds for a More Creative Society. *Proc. Learning & Leading with Technology.*

Resnick, M. (2013). *Lifelong Kindergarten. Cultures of Creativity.* LEGO Education.

Resnick, M. & Rusk, N. (1996). The Computer Clubhouse: Preparing for life in a digital world. *IBM Systems Journal, 35*(3.4), 431–439. doi:10.1147/sj.353.0431

Resnick, M. & Silverman, B. (2005). Some reflections on designing construction kits for kids. In *Proceedings of the 2005 conference on Interaction design and children* (pp. 117–122).

Resnick, M., Silverman, B., Kafai, Y., Maloney, J., Monroy-Hernández, A., Rusk, N. et al. (2009). Scratch: programming for all. *Communications of the ACM, 52*(11), 60. doi:10.1145/1592761.1592779

Richter, T. & Naumann, J. (2010). *Inventar zur Computerbildung (INCOBI-R) – Eine revidierte Fassung des Inventars zur Computerbildung (INCOBI-R).*

Richter, T., Naumann, J. & Groeben, N. (2001). Das Inventar zur Computerbildung (INCOB): Ein Instrument zur Erfassung von Computer Literacy und computerbezogenen Einstellungen bei Studierenden der Geistes- und Sozialwissenschaften. *Psychologie in Erziehung und Unterricht, 48.*

Riese, J. (2009). *Professionelles Wissen und professionelle Handlungskompetenz von (angehenden) Physiklehrkräften.* Berlin: Logos-Verlag.

Rogozhkina, I. & Kushnirenko, A. (2011). PiktoMir: teaching programming concepts to preschoolers with a new tutorial environment. *Procedia – Social and Behavioral Sciences, 28*, 601–605. doi:10.1016/j.sbspro.2011.11.114

Romeike, R. (2008). *Kreativität im Informatikunterricht.* Potsdam: Universität Potsdam.

Romeike, R. & Reichert, D. (2011). PicoCrickets als Zugang zur Informatik in der Grundschule. Presented at *Informatik in Bildung und Beruf–INFOS 2011–14. GI-Fachtagung Informatik und Schule.*

Ropohl, G. (1999). *Allgemeine Technologie: eine Systemtheorie der Technik* (2nd ed.). München/Wien: Hanser.

Rosen, L. D., Lim, A. F., Felt, J., Carrier, L. M., Cheever, N. A., Lara-Ruiz, J. M. et al. (2014). Media and technology use predicts ill-being among children, preteens and teenagers independent of the negative health impacts of exercise and eating habits. *Computers in Human Behavior, 35*, 364–375. doi:10.1016/j.chb.2014.01.036

Rozenszajn, R. & Yarden, A. (2014). Expansion of Biology Teachers' Pedagogical Content Knowledge (PCK) During a Long-Term Professional Development Program. *Research in Science Education, 44*(1), 189–213. doi:10.1007/s11165-013-9378-6

Ruf, A., Mühling, A. & Hubwieser, P. (2014). Scratch vs. Karel: impact on learning outcomes and motivation. In *Proceedings of the 9th Workshop in Primary and Secondary Computing Education* (pp. 50–59).

Rushkoff, D. (2010). *Program or be programmed: ten commands for a digital age.* New York: Or Books.

Sachser, N. (2004). Neugier, Spiel und Lernen: Verhaltensbiologische Anmerkungen zur Kindheit. *Zeitschrift für Pädagogik 50*(4), 475–486.

Saeli, M. (2012). *Teaching Programming for Secondary School: A Pedagogical Content Knowledge Based Approach.* Eindhoven: Technische Universiteit Eindhoven.

Schefe, P. (1999). Softwaretechnik und Erkenntnistheorie. *Informatik-Spektrum, 22*(2), 122–135. doi:10.1007/s002870050131

Schiefele, U. (2009). Motivation. In E. Wild & J. Möller (Eds.), *Pädagogische Psychologie* (pp. 151–177). Berlin, Heidelberg: Springer Berlin Heidelberg.

Schneider, W., Körkel, J. & Weinert, F. E. (1989). Domain-specific knowledge and memory performance: A comparison of high- and low-aptitude children. *Journal of Educational Psychology, 81*(3), 306–312. doi:10.1037/0022-0663.81.3.306

Schubert, S. & Schwill, A. (2011). *Didaktik der Informatik* (2nd ed.). Heidelberg: Spektrum, Akademischer Verlag.

Schulte, C. (2001). Vom Modellieren zum Gestalten: Objektorientierung als Impuls für einen neuen Informatikunterricht? *informatica didacta Zeitschrift für fachdidaktische Grundlagen der Informatik, 3.*

Schulte, C. (2003). *Lehr-Lernprozesse im Informatik-Anfangsunterricht theoriegeleitete Entwicklung und Evaluation eines Unterrichtskonzepts zur Objektorientierung in der Sekundarstufe II* (Doctoral dissertation). Paderborn, Univ. Retrieved March, 2022, from http://ubdata.uni-paderborn.de/ediss/17/2003/schulte/disserta.pdf

Schulte, C. (2008a). Interesse wecken und Grundkenntnisse vermitteln. *3. Münsteraner Workshop zur Schulinformatik.* Münster.

Schulte, C. (2008b). Duality Reconstruction – Teaching Digital Artifacts from a Socio-technical Perspective. In R.T. Mittermeir & M.M. Sysło (Eds.), *Informatics Education – Supporting Computational Thinking* (Vol. 5090, pp. 110–121). Berlin, Heidelberg: Springer Berlin Heidelberg.

Schulte, C. (2009). Dualitätsrekonstruktion als Hilfsmittel zur Entwicklung und Planung von Informatikunterricht. *Zukunft braucht Herkunft: 25 Jahre "INFOS – Informatik und Schule"* (pp. 355–366). Presented at INFOS, Berlin.

Schulte, C. (2012). Uncovering structure behind function: the experiment as teaching method in computer science education. In *Proceedings of the 7th Workshop in Primary and Secondary Computing Education* (pp. 40–47).

Schulte, C. (2013). Reflections on the role of programming in primary and secondary computing education. In *Proceedings of the 8th Workshop in Primary and Secondary Computing Education* (pp. 17–24).

Schulte, C. & Bennedsen, J. (2006). What do teachers teach in introductory programming? In *Proceedings of the second international workshop on Computing education research* (pp. 17–28).

Schulte, C. & Knobelsdorf, M. (2011). Medien nutzen, Medien gestalten – eine qualitative Analyse der Computernutzung. In C. Albers, J. Magenheim & D.M. Meister (Eds.), *Schule in der digitalen Welt* (pp. 97–115). Wiesbaden: VS Verlag für Sozialwissenschaften.

Schulz, W. (1997). Die lehrtheoretische Didaktik. In H. Gudjons (Ed.), *Didaktische Theorien* (pp. 35–56). Hamburg: Bergmann + Helbig.

Schwarzkopf, H. & Zolg, M. (1997). Kann der Computer denken? Gespräche mit Kindern. *Die Grundschulzeitschrift, 11*(108), 44–46.

Schwill, A. (1993). Fundamentale Ideen der Informatik. *Zentralblatt für Didaktik der Mathematik, 25*(1), 20–31.

Schwill, A. (1995). Fundamentale Ideen in Mathematik und Informatik. *Bericht über die 12. Arbeitstagung des AK "Mathematikunterricht und Informatik" der GDM* (pp. 18–25).

Schwill, A. (2001). Ab wann kann man mit Kindern Informatik machen. *INFOS2001-9. GI-Fachtagung Informatik und Schule GI-Edition*, 13–30.

Seel, N. M. (2003). *Psychologie des Lernens: Lehrbuch für Pädagogen und Psychologen*. UTB Pädagogik, Psychologie: Vol. 8198.

Senatsverwaltung für Bildung, Jugend und Wissenschaft Berlin (Ed.). (2014). *Berliner Bildungsprogramm für Kitas und Kindertagespflege*. Weimar: verlag das netz.

Shinners-Kennedy, D. & Fincher, S. A. (2013). Identifying threshold concepts: from dead end to a new direction. In *Proceedings of the ninth annual international ACM conference on International computing education research* (pp. 9-18).

Shulman, L. S. (1986). Those Who Understand: Knowledge Growth. *Teaching Educational, 15*(2), 4–14.

Shulman, L. S. (1987). Knowledge and teaching: Foundations of the new reform. *Harvard Educational Review, 57*(1), 1–21.

Siemens, G. (2008). Connectivism: A Learning Theory of the Digital Age 2008. *International Journal of Instructional Technology & Distance Learning*.

Sodian, B. (2002). Die Entwicklung des bereichsspezifischen Wissens. In R. Oerter & L. Montada (Eds.), *Entwicklungspsychologie*. Weinheim: Psychologie Verlags Union.

Sodian, B., Koerber, S. & Thoermer, C. (2006). Zur Entwicklung des naturwissenschaftlichen Denkens im Vor- und Grundschulalter. In P. Nentwig & S. Schanze (Eds.), *Es ist nie zu früh. Naturwissenschaftliche Bildung in jungen Jahren* (pp. 11–20). Münster: Waxmann.

Spiro, R. J., Feltovich, P. J., Jacobson, M. J. & Coulson, R. L. (1991). Cognitive flexibility, constructivism and hypertext: Random access instruction for advanced knowledge acquisition in ill-structured domains. *Educational Technology*, 24–33.

Stechert, P. (2009). *Fachdidaktische Diskussion von Informatiksystemen und der Kompetenzentwicklung im Informatikunterricht* (Vol. 2). Universitätsverlag Potsdam.

Steinbuch, K. (1957). Automatische Informationsverarbeitung. *SEG-Nachrichten (Technische Mitteilungen der Standard Elektrik Gruppe)–Firmenzeitschrift*, 4, p. 171.

Stern, E. (2002). Wie abstrakt lernt das Grundschulkind? In H. Petillon (Ed.), *Individuelles und soziales Lernen in der Grundschule* (pp. 27–42). Wiesbaden: VS Verlag für Sozialwissenschaften. doi:10.1007/978-3-322-99278-9_2

Stiftung Haus der kleinen Forscher. (2013). *Kommst du mit die Zeit entdecken? Ideen zum Forschen und Staunen rund um das Phänomen "Zeit"*. Berlin: Stiftung Haus der kleinen Forscher.

Stiftung Haus der kleinen Forscher. (2014). *Kannst du mich verstehen? Die Vielfalt der Kommunikation erkunden und erforschen*. Berlin: Stiftung Haus der kleinen Forscher.

Stiftung Haus der kleinen Forscher. (2019). *Pädagogischer Ansatz der Stiftung "Haus der kleinen Forscher"* (6th, completely revised edition). Berlin: Stiftung Haus der kleinen Forscher. Available at haus-der-kleinen-forscher.de/de/fortbildungen

Taub, R., Ben-Ari, M. & Armoni, M. (2009). The effect of CS unplugged on middle-school students' views of CS. *ACM SIGCSE Bulletin, 41*(3), 99–103.

Tedre, M. & Apiola, M. (2013). Three computing traditions in school computing education. In *Improving computer science education* (pp. 108-124). Routledge.

The Royal Society. (2012). *Shut down or restart? The way forward for computing in UK schools*. The Royal Society.

Tillmann, A., Fleischer, S. & Hugger, K.-U. (Eds.). (2014). *Handbuch Kinder und Medien* (Digitale Kultur und Kommunikation). Wiesbaden: Springer VS.

UNESCO. (2012). *UNESCO ICT Competency Framework for Teachers*. Available at http://unesdoc.unesco.org/images/0021/002134/213475E.pdf

van Lück, W. (1986). *Informations- und kommunikationstechnische Grundbildung und der Computer als Medium im Fachunterricht*. Soest: Soester Verlagskontor.

Van Merrienboer, J. J. G. & Krammer, H. P. M. (1987). Instructional strategies and tactics for the design of introductory computer programming courses in high school. *Instructional Science, 16*(3), 251–285. doi:10.1007/BF00120253

Vekiri, I. & Chronaki, A. (2008). Gender issues in technology use: Perceived social support, computer self-efficacy and value beliefs, and computer use beyond school. *Computers & Education, 51*(3), 1392–1404. doi:10.1016/j.compedu.2008.01.003

Walter-Herrmann, J. & Büching, C. (Eds.). (2013). *FabLab of machines, makers and inventors* (Cultural and media studies). Bielefeld: transcript.

Wegner, P. (1997). Why interaction is more powerful than algorithms. *Communications of the ACM, 40*(5), 80–91. doi:10.1145/253769.253801

Weinert, F. E. (2001). Concept of Competence: A conceptual clarification. In D.S. Rychen & L.H. Salganik (Eds.), *Defining and Selecting Key Competencies*. Seattle: Hogrefe and Huber.

Weintrop, D. & Wilensky, U. (2015). Using Commutative Assessments to Compare Conceptual Understanding in Blocks-based and Text-based Programs. In *Proceedings of the eleventh annual international conference on international computing education research* (pp. 101–110).

Wenger, E. (1998). Communities of Practice: Learning as a social system. *Systems thinker, 9*(5), 2–3.

Wiener, N. (1948). *Kybernetik – Regelung und Nachrichtenübertragung in Lebewesen und in der Maschine*. Düsseldorf: Econ-Verlag Düsseldorf.

Wiesner, A. (2008). *Lerngruppe – Logisches Denken*. München: Ars-Ed.

Wiesner, B. & Brinda, T. (2007). Erfahrungen bei der Vermittlung algorithmischer Grundstrukturen im Informatikunterricht der Realschule mit einem Robotersystem. *Didaktik der Informatik in Theorie und Praxis*, 113–124.

Wikipedia. (2007). Finite state machine. *Wikipedia*. Retrieved March, 2022, from https://commons.wikimedia.org/wiki/File:Finite_state_machine_example_with_comments.svg

Wikipedia. (2011). Petri-Netz. *Wikipedia*. Retrieved March, 2022, from https://de.wikipedia.org/w/index.php?title=Petri-Netz&oldid=155851849#/media/Datei:Seasons_1.svg

Wikipedia. (2014). Artefakt. *Wikipedia*. Retrieved March, 2022, from https://de.wikipedia.org/w/index.php?title=Artefakt&stableid=136435468

Wikipedia. (2015). UML Sequenzdiagramm. Wikipedia. Retrieved March, 2022, from https://de.wikipedia.org/wiki/Datei:UmlSequenzdiagramm-2.svg

Wikipedia. (2016). Softwarequalität. Wikipedia. Retrieved March, 2022, from https://de.wikipedia.org/wiki/Softwarequalit%C3%A4t

Wikipedia. (2016b). WYSIWYG. *Wikipedia*. Retrieved March, 2022, from https://de.wikipedia.org/w/index.php?title=WYSIWYG&oldid=157393007

Williams, J. & Lockley, J. (2012). Using CoRes to Develop the Pedagogical Content Knowledge (PCK) of Early Career Science and Technology Teachers. *Journal of Technology Education, 24*(1). doi:10.21061/jte.v24i1.a.3

Wilson, C. (2015). Hour of code – a record year for computer science. *ACM Inroads, 6*(1), 22–22. doi:10.1145/2723168

Wilson, C., Sudal, L. A., Stephenson, C. & Stehlik, M. (2010). *Running on Empty: The Failure to Teach Computer Science in the Digital Age. Association for Computing Machinery and the Computer Science Teachers Association*. New York.

Wing, J. M. (2006). Computational thinking. *Communications of the ACM, 49*(3), 33–35. doi:10.1145/1118178.1118215

Wing, J. M. (2008). Computational thinking and thinking about computing. *Philosophical Transactions of the Royal Society A: Mathematical, Physical and Engineering Sciences, 366*(1881), 3717–3725. doi:10.1098/rsta.2008.0118

Xie, L., Antle, A. N. & Motamedi, N. (2008). Are tangibles more fun?: comparing children's enjoyment and engagement using physical, graphical and tangible user interfaces. In *Proceedings of the 2nd international conference on tangible and embedded interaction* (pp. 191–198).

Yardi, S. & Bruckman, A. (2007). What is computing? Bridging the gap between teenagers' perceptions and graduate students' experiences. In *Proceedings of the third international workshop on computing education research* (pp. 39–50).

Ziegenbalg, J. (1985). *Programmieren lernen mit Logo*. München: Hanser.

Zolg, M. (2006). Mut zur Technik! *Weltwissenschaft Sachunterricht, 4*.

## Conclusion and Outlook –

"Haus der kleinen Forscher" Foundation

Brünger, K., Franke-Wiekhorst, A., Griffiths, K., Günther, C. & Radtke, M. (2019). Informatische Bildung für Kinder im Kita- und Grundschulalter – ein Konzept zum entdeckenden und forschenden Lernen für die Praxis. *GDSU-Journal, 9*, 106–117.

Chaudron, S. (2015). *Young children (0-8) and digital technology: a qualitative exploratory study across seven countries*. (Joint Research Centre – Institute for the Protection and Security of the Citizen, Ed.). Luxembourg: Publications Office of the European Union.

Ferreira, J.-A., Ryan, L. & Tilbury, D. (2006). *Whole-School Approaches to Sustainability: A review of models for professional development in pre-service teacher education*. Canberra: Australian Government Department of the Environment and Heritage and the Australian Research Institute in Education for Sustainability (ARIES).

GI – Gesellschaft für Informatik e.V. (2008). *Bildungsstandards Informatik für die Sekundarstufe I: Empfehlungen der Gesellschaft für Informatik e.V.* (Vol. 28). LOG IN.

"Haus der kleinen Forscher" Foundation. (2018). Early Science Education – Goals and Process-Related Quality Criteria for Science Teaching (Scientific Studies on the Work of the "Haus der kleinen Forscher" Foundation, Vol. 5). Opladen, Berlin, Toronto: Verlag Barbara Budrich. Available at: haus-der-kleinen-forscher.de/de/wissenschaftliche-begleitung

Marquardt-Mau, B. (2004). Ansätze zur Scientific Literacy. Neue Wege für den Sachunterricht. In A. Kaiser & D. Pech (Eds.), *Neuere Konzeptionen und Zielsetzungen im Sachunterricht* (pp. 67–83). Hohengehren: Schneider Verlag.

Stiftung Haus der kleinen Forscher. (2013). *Der Forschungskreis*. Berlin: Stiftung Haus der kleinen Forscher. Available at: haus-der-kleinen-forscher.de/de/praxisanregungen

Stiftung Haus der kleinen Forscher (Ed.). (2015). *Wissenschaftliche Untersuchungen zur Arbeit der Stiftung "Haus der kleinen Forscher"* (Vol. 7). Schaffhausen: SCHUBI Lernmedien AG. Available at: haus-der-kleinen-forscher.de/de/wissenschaftliche-begleitung

Stiftung Haus der kleinen Forscher. (2016). *Der Mathematikkreis*. Berlin: Stiftung Haus der kleinen Forscher. Available at: haus-der-kleinen-forscher.de/de/praxisanregungen

Stiftung Haus der kleinen Forscher (Ed.). (2017a). *Frühe mathematische Bildung – Ziele und Gelingensbedingungen für den Elementar- und Primarbereich* (Wissenschaftliche Untersuchungen zur Arbeit der Stiftung "Haus der kleinen Forscher", Vol. 8). Opladen, Berlin, Toronto: Verlag Barbara Budrich. Available at: haus-der-kleinen-forscher.de/de/wissenschaftliche-begleitung

Stiftung Haus der kleinen Forscher. (2017b). *Wie nutzen Erzieherinnen und Erzieher digitale Geräte in Kitas? – Eine repräsentative Telefonumfrage.* Berlin: Stiftung Haus der kleinen Forscher. Available at: haus-der-kleinen-forscher.de/de/wissenschaftliche-begleitung

Stiftung Haus der kleinen Forscher. (2017c). *Informatik entdecken – mit und ohne Computer.* Berlin: Stiftung Haus der kleinen Forscher. Available at: haus-der-kleinen-forscher.de/de/praxisanregungen

Stiftung Haus der kleinen Forscher. (2017d). *Materialpaket Informatik entdecken – mit und ohne Computer.* Berlin: Stiftung Haus der kleinen Forscher.

Stiftung Haus der kleinen Forscher. (2018). *Technikkreis.* Berlin: Stiftung Haus der kleinen Forscher. Available at: haus-der-kleinen-forscher.de/de/praxisanregungen

Stiftung Haus der kleinen Forscher. (2019a). *Pädagogischer Ansatz der Stiftung „Haus der kleinen Forscher"* (6th, completely revised edition). Berlin: Stiftung Haus der kleinen Forscher. Available at: haus-der-kleinen-forscher.de/de/fortbildungen

Stiftung Haus der kleinen Forscher (Ed.). (2019b). *Frühe Bildung für nachhaltige Entwicklung – Ziele und Gelingensbedingungen* (Wissenschaftliche Untersuchungen zur Arbeit der Stiftung "Haus der kleinen Forscher", Vol. 12). Opladen, Berlin, Toronto: Verlag Barbara Budrich. Available at: haus-der-kleinen-forscher.de/de/wissenschaftliche-begleitung

Stiftung Haus der kleinen Forscher. (2020a). *MINT geht digital – Entdecken und Forschen mit digitalen Medien.* Berlin: Stiftung Haus der kleinen Forscher. Available at: haus-der-kleinen-forscher.de/de/praxisanregungen

Stiftung Haus der kleinen Forscher. (2020b). *Monitoring-Bericht 2018/2019 der Stiftung "Haus der kleinen Forscher".* Berlin: Stiftung Haus der kleinen Forscher. Available at: haus-der-kleinen-forscher.de/de/wissenschaftliche-begleitung

Stiftung Haus der kleinen Forscher. (2021). *Digitale Bildung – Chance für gute frühe MINT-Bildung für nachhaltige Entwicklung. Positionspapier der Stiftung "Haus der kleinen Forscher"*. Berlin: Stiftung Haus der kleinen Forscher. Available at: haus-der-kleinen-forscher.de/de/ueberuns/presse/pressemitteilungen

Stiftung Haus der kleinen Forscher. (2022). *Monitoring-Bericht 2020/2021 der Stiftung "Haus der kleinen Forscher"*. Berlin: Stiftung Haus der kleinen Forscher. Available at: haus-der-kleinen-forscher.de/de/wissenschaftliche-begleitung

# Appendix

# Appendix

**Mapping the components of international standards and curricula into the framework of a competence model for computer science education at the primary level**

In the following, the original formulations of the competencies of the international standards and curricula are assigned to the Content Domains of the competence model discussed in Chapter 3 in order to substantiate the interpretation by the team of authors, on the one hand, and to clarify the emphases made in the international approaches, on the other.

In this context, the Process Domains are to be concluded rather indirectly, as explained in Section 2.5.2, while the Content Domains are explicitly stated.

## (C1) Information and Data

*CAS (Computing at School; Great Britain)*
Key Stage 1 (pre-school – grade 2, age 5-7)

- A pupil should **understand** how computers **represent data**. (CAS, p. 16, Data)
- **Information** can be **stored** and **communicated** in a variety of forms e.g., numbers, text, sound, image, video. (CAS, p. 16, Data)
- Computers use **binary** switches (on/off) to store information. (CAS, p. 16, Data)
- Binary (yes/no) answers can directly provide useful information (e.g., present or absent), and be **used for decision**. (CAS, p. 16, Data)

Key Stage 2 (grades 3-6, age 7-11)

- Similar information can be represented in multiple ways. (CAS, p. 16, Data)
- Introduction to **binary representation** [representing **names**, **objects** or **ideas** as sequences of 0s and 1s]. (CAS, p. 16, Data)
- Difference between **data** and **information**. (CAS, p. 16, Data)
- Structured data can be stored in **tables** with rows and columns. Data in tables can be **sorted**. Tables can be **searched** to answer **questions**. Searches can use one or more columns of the table. (CAS, p. 16, Data)
- Data may contain errors and this affects the search results and decisions based on the data. **Errors** may be reduced using **verification** and **validation**. (CAS, p. 16, Data)

- A pupil should understand the principles underlying how **data** is transported on the **Internet**. (CAS, p.18, Communication and the Internet)

*CSTA (Computer Science Teachers Association; USA)*
Level 1: (pre-school – grade 3, age 5-8)

- **Demonstrate** how 0s and 1s can be used to **represent information**. (CSTA p. 13, L1:3.CT 5)
    - Process Domains P5 Representing & Interpreting, possibly also P1 Modelling or P2 Reasoning or P3 Structuring

Level 2: (grades 3-6, age 8-11)

- Demonstrate how a string of bits can be used to **represent alphanumeric information**. (CSTA p. 13, L1:6. CT 3.)
    - especially Process Domain P5 Representing & Interpreting
- **Gather** and **manipulate** data using a variety of digital **tools**. (CSTA p.14, L1: 6. CPP 10.)
    - Process Domains P5 Representing & Interpreting and P1 Modelling

*New Zealand*
Level 1 (grades 1-3, age 5-7)

- Conservative: not included
- Advanced: How 0s and 1s represent information; patterns and symbols; pixels and file size.
    - P5 (Interpreting)

Level 2 (grades 3-5, age 7-9)

- How two different symbols can **represent information** (e.g., binary numbers).
    - P5 (present examples, interpret representations), possibly P1 (find suitable representation for problem => modelling), possibly P2 (select from possible representations for problem => justify, evaluate)
- **Representation** of **text** and **images** using binary; codes and symbols.
    - P1 or P5, possibly P2 and P3

*Swiss curriculum 21*
Learners can **represent** (P1, P5), **structure** (P3) and **evaluate** (P5) **data** from their environment. (MI.2.1, p.14)

Grades 1-2, age 6-7:

- Learners can **order** things according to properties they have chosen themselves (P3) so that they can **find** an object with a certain property more quickly (P5) (e.g., colour, shape, size).

Grades 3-4, age 8-9:

- Learners can use different **ways of presenting data** (P5) (e.g., symbols, tables, graphs).
- Learners can encrypt data using **secret scripts they have developed themselves** (P5).
- Learners know **analogue** and **digital** representations of data (text, number, image and sound) and can **match** the corresponding file types (P3).

Grades 4-6, age 9-11:

- Learners **recognise** and **use** (P1, P3, P5) **tree and network structures** (e.g., folder structure on computer, family tree, mind map, website).
- Learners **understand** (P2, P5) how **error-detecting** and **error-correcting codes** work.

## (C2) Algorithms (and Programming)

*CAS*
Key Stage 1 (pre-school – grade 2, age 5-7)

- Algorithms are **sets of instructions** for achieving goals, made up of **pre-defined steps** [the 'how to' part of a recipe for a cake]. (CAS, p. 13, Algorithms)
- Algorithms can be **represented in simple formats** [storyboards and narrative text]. (CAS, p. 13, Algorithms)
- They can describe **everyday activities** and can be **followed by humans** and **by computers**. (CAS, p. 13, Algorithms)
- A pupil should **understand what an algorithm is**, and what algorithms can be **used for**. (CAS, p. 13, Algorithms)
- **Steps** can be **repeated** and some steps can be **made up of smaller steps**. (CAS, p. 13, Algorithms)

Key Stage 2 (grades 3-6, age 7-11)

- Algorithms can be **represented symbolically** [flowcharts] or using instructions in a clearly defined **language** [turtle graphics]. (CAS, p. 13, Algorithms)
- Algorithms can include **selection (if)** and **repetition (loops)**. (CAS, p. 13, Algorithms)
- Algorithms may be **decomposed** into component parts (**procedures**), each of which itself contains an algorithm. (CAS, p. 13, Algorithms)
- Algorithms should be stated **without ambiguity** and care and **precision** are necessary to avoid errors. (CAS, p. 13, Algorithms)
- Programs can include **repeated instructions**. (CAS, p. 14, Programs)
- A computer **program** is a sequence of instructions written to perform a specified **task** with a computer. (CAS, p. 14, Programs)
- The idea of a program as a **sequence** of statements written in a **programming language** [Scratch]. (CAS, p. 14, Programs)
- One or more mechanisms for **selecting** which **statement sequence** will be executed, **based upon** the **value** of some data item. (CAS, p. 14, Programs)
- One or more mechanisms for **repeating** the execution of a sequence of statements, and using the **value** of some data item **to control the number** of times the sequence is repeated. (CAS, p. 14, Programs)
- Programs can be created using **visual tools**. Programs can work with different types of data. They can use a **variety of control structures** [selections and procedures]. (CAS, p. 15, Programs)
- Programs are **unambiguous** and that care and precision is necessary to avoid errors. (CAS, p. 15, Programs)
- Programs are **developed according to a plan** and then **tested**. Programs are corrected if they fail these tests. (CAS, p. 15, Programs)
- The **behaviour** of a program should be **planned**. (CAS, p. 15, Programs)
- A well-written program tells a reader the story of how it works, both in the code and in **human-readable comments**. (CAS, p. 15, Programs)
- A web page is an **HTML** script that constructs the **visual appearance**. It is also the carrier for other **code** that can be **processed** by the browser. (CAS, p. 15, Programs)

- Computers can be **programmed** so they appear to **respond 'intelligently' to certain inputs**. (CAS, p. 15, Programs)
- The difference between **constants** and **variables** in programs. (CAS, p. 16, Data)

*CSTA*
Level 1: (pre-school – grade 3, age 5-8)

- **Use technology resources** (e.g., puzzles, logical thinking programs) to solve age-appropriate problems. (CSTA p. 13, L1:3.CT 1.)
    - P0, P1, possibly P3 and P5
- Understand how to **arrange (sort) information** into useful order, such as sorting students by birth date, without using a computer. (CSTA p. 13, L1:3.CT 3.)
    - P0, P1, P3, P5; possibly P2 and P4
- **Construct** a set of **statements** (P1) to be acted out to accomplish a simple task (e.g., turtle instructions). (CSTA p.14, L1: 3. CPP 4.)

Level 2: (grades 3-6, age 8-11)

- Understand and use the **basic steps in algorithmic problem-solving** (e.g., problem statement and exploration, examination of sample instances, design, implementation, and testing). (CSTA p. 13, L1:6.CT 1.)
    - P0, P1, P5, possibly P2, P3, P4
- Develop a **simple understanding of an algorithm** (P1, (e.g., search, sequence of events, or sorting) using computer-free exercises. (CSTA p. 13, L1:6. CT 2.)
    - P1, P5, possibly P2, P3, P4
- Make a **list of sub-problems** (P3, P1) to consider while addressing a larger problem. (CSTA p. 13, L1:6. CT 5.)
- **Construct a program** (P1, P5) as a set of **step-by-step instructions** to be acted out (e.g., make a peanut butter and jelly sandwich activity). (CSTA p.14, L1: 6. CPP 5.)

*New Zealand*
Level 1 (grades 1-3, age 5-7)

- Understand **what algorithms are** (P2, P5) and **follow an algorithm** (P5).
- **Sorting** (P1, P5) and **patterns** (P3).

- **Create** and **debug** (P1, P5, possibly P3) simple programs (e.g., turtle instructions) with simple **sequencing** and **repetition.**

Level 2 (grades 3-5, age 7-9)

- **Decompose** (P1, P5) problems into steps.
- **Explain** (P5, possibly P2) and **correct** (P1) **errors** in algorithms.
- Block-based/visual programming including simple **iteration**.
- design, implement, test and debug (P1, incl. P3 and P5, possibly P2 and P4) an interactive application in a visual programming language (sequence and **selection; input**).

*Swiss curriculum 21*

Learners can analyse simple problems (P2/P3/P5), describe possible solution procedures (P5) and implement them in programmes (P1).

Grades 1-2, age 6-7:

- Learners can **recognise and follow formal instructions** (P5, AFB I-II) (e.g., cooking and baking recipes, instructions for games and handicrafts, dance choreographies).

Grades 3-4, age 8-9:

- Learners can find **solutions** (P1, P5) to simple problems **by trial and error** and **check** them **for correctness** (P1, P5) (e.g., find a way, develop a game strategy). They can **compare** different **ways of solving problems** (P2).
- Learners can **recognise processes** with **loops** and **branches** from their environment **(P3)**, describe and **present them in a structured way (P5)** (e.g., by means of flow charts).
- Learners can **read** and **manually execute** simple sequences with loops, conditional statements and parameters **(P5, possibly P4)**.
- Learners **understand (P2, possibly P5)** that a computer can **only** execute **predefined instructions** and that a programme is a sequence of such instructions.
- Learners can write and test **programmes** with loops, conditional instructions and parameters **(P1+ Testing)**.

## (C3) Languages and Automata

*CAS*

Key Stage 1 (pre-school – grade 2, age 5-7)

- Computers need more **precise instructions** than humans do. (CAS, p. 13, Algorithms)
- Computers (understood here to include all devices controlled by a processor, thus including programmable toys, phones, game consoles and PCs) are controlled by **sequences of instructions**. (CAS, p. 14, Programs)
- A computer program is like the narrative part of a story, and the computer's job is to do what the narrator says. Computers have **no intelligence,** and so **follow** the narrator's **instructions blindly**. (CAS, p. 14, Programs)

Key Stage 2 (grades 3-6, age 7-11)

- Programs can model and **simulate environments** to answer **"What if"** questions. (CAS, p. 14, Programs)
    - This could also be represented in the form of automata.
- Programs are **unambiguous** and that care and precision is necessary to avoid errors. (CAS, p. 15, Programs)
- A web page is an **HTML** script that constructs the **visual appearance**. It is also the carrier for other **code** that can be **processed** by the browser. (CAS, p. 15, Programs)
- The format of URLs. (CAS, p. 19, Communication and the Internet)

*CSTA*

Level 1: (pre-school – grade 3, age 5-8)

- Construct a set of statements to be acted out to accomplish a simple task (e.g., **turtle instructions**). (CSTA p.14, L1: 3. CPP 4.)

*New Zealand*

Level 1 (grades 1-3, age 5-7)

- Nothing

Level 2 (grades 3-5, age 7-9)

- Nothing

*Swiss curriculum 21*
Grades 3-4, age 8-9:

- Learners **understand** (P2, possibly P5) that a computer can **only** execute **pre-defined instructions** and that a programme is a sequence of such instructions.

## (C4) Informatics Systems
*CAS*
Key Stage 1 (pre-school– grade 2, age 5-7)

- Particular **tasks** can be accomplished by creating a program for a computer. Some computers allow their users to create their own programs. (CAS, p. 14, Programs)
- Computers typically accept **inputs,** follow a stored sequence of **instructions** and produce **outputs**. (CAS, p. 14, Programs)
- A pupil should know the main **components** that make up a **computer** system, and how they fit together (their **architecture**). (CAS, p. 17, Computers)
- **Computers** are electronic devices using stored sequences of instructions. (CAS, p. 17, Computers)
- Computers typically accept input and produce outputs, with examples of each in the context of PCs. (CAS, p. 17, Computers)
- Many devices now contain computers. (CAS, p. 17, Computers)
- Web **browser** is a program used to view pages. (CAS, p.18, Communication and the Internet)

Key Stage 2 (grades 3-6, age 7-11)

- Computers are devices for executing programs. (CAS, p. 17, Computers)
- Application software is a computer program designed to perform user tasks. (CAS, p. 17, Computers)
- The **operating system** is a software that manages the relationship between the application software and the hardware. (CAS, p. 17, Computers)
- Computers consist of a number of **hardware components** each with a specific role [e.g., CPU, Memory, hard disk, mouse, monitor]. (CAS, p. 17, Computers)
- Both the operating system and application software **store data** (e.g., in memory and a file system). (CAS, p. 17, Computers)

- The above applies to devices with embedded computers (e.g., digital cameras), handheld technology (e.g. smart phones) and personal computers. (CAS, p. 18, Computers)
- A **variety of operating systems** and application software is typically available for the same hardware. (CAS, p. 18, Computers)
- Users can **prevent or fix problems** that occur with computers (e.g., connecting hardware, protection against viruses). (CAS, p. 18, Computers)
- A pupil should understand the principles underlying how **data** is transported on the **Internet**. (CAS, p.18, Communication and the Internet)
- The **Internet** is a collection of computers connected together sharing the same way of communicating. The Internet is not the web, and the web is not the Internet. (CAS, p.18, Communication and the Internet)
- These connections can be made using a **range of technologies** (e.g., network cables, telephone lines, WiFi, mobile signals, carrier pigeons). (CAS, p. 19, Communication and the Internet)
- The Internet supports multiple **services** (e.g., the Web, e-mail, VoIP). (CAS, p. 19, Communication and the Internet)
- The relationship between web **servers**, web **browsers**, **websites** and web **pages**. (CAS, p. 19, Communication and the Internet)
- The role of **search engines** in allowing users to find specific web pages and a basic understanding of how results may be ranked. (CAS, p. 19, Communication and the Internet)
- Issues of **safety and security** from a technical perspective. (CAS, p. 19, Communication and the Internet)

*CSTA*
Level 1: (pre-school – grade 3, age 5-8)

- **Use** writing **tools**, digital **cameras**, and **drawing** tools to illustrate thoughts, ideas, and stories in a step-by-step manner. (CSTA p.13, L1:3.CT 2.)
- **Recognise** that software is created to control computer operations. (CSTA p.13, L1:3. CT 4.)
- Use standard **input and output devices** to successfully operate computers and related technologies. (CSTA p.14, L1: 3. CD 1.)
- Use technology resources to conduct age-appropriate **research**. (CSTA p.14, L1: 3. CPP 1.)

- Use developmentally appropriate multimedia resources (e.g., interactive books and educational software) to support **learning** across the curriculum. (CSTA p.14, L1: 3. CPP 2.)

- Create developmentally appropriate **multimedia products** with support from teachers, family members, or student partners. (CSTA p.14, L1: 3. CPP 3.)

Level 2: (grades 3-6, age 8-11)

- Demonstrate an appropriate level of proficiency with **keyboards** and other **input** and **output devices**. (CSTA p.14, L1: 6. CD 1.)

- Apply strategies for **identifying** simple **hardware and software problems** that may occur during use. (CSTA p.14, L1: 6. CT 3.)

- Identify that **information** is coming to the computer from many **sources over a network**. (CSTA p.14, L1: 6. CT 4.)

- Describe how a **simulation** can be used to **solve a problem**. (CSTA p.13, L1:6. CT 4.)

- Use technology resources (e.g., calculators, data collection probes, mobile devices, videos, educational software, and web tools) for **problem-solving** and **self-directed learning**. (CSTA p.14, L1: 6. CPP 1.)

- Use general-purpose productivity tools and peripherals to support personal productivity, remediate skill deficits, and facilitate learning. (CSTA p.14, L1: 6. CPP 2.)

- Use technology tools (e.g., multimedia and text authoring, presentation, web tools, digital cameras, and scanners) for individual and collaborative **writing, communication,** and **publishing** activities. (CSTA p.14, L1: 6. CPP 3.)

- Use computing devices to **access** remote information, **communicate** with others in support of direct and **independent learning,** and pursue personal interests. (CSTA p.14, L1: 6. CPP 7.)

- **Navigate** between webpages using hyperlinks and conduct simple **searches** using search engines. (CSTA p.14, L1: 6. CPP 8.)

*New Zealand*
Level 1 (grades 1-3, age 5-7)

- **Use** (P0) common input/output **devices** (e.g., keyboard, pointing device, touch screen)

- **Describe** (P3) hardware and software components; input/output devices

- Create, organise, store, manipulate and retrieve **content** (P0) including **multi-media** (P0)

Level 2 (grades 3-5, age 7-9)

- Peripheral devices; data capture; data transfer (P0)
- Identify simple **software problems** (P3, P5)
- Use **search** technology (P0)
- **Select, use and combine software** (P3, possibly P2) to collect, organise and present **data**; simple **spreadsheets** and **charts**

*Swiss curriculum 21*
Learners understand the **structure** and **functioning** of information processing concepts and apply concepts of **secure data processing**.

Grades 1-2, age 6-7:

- Learners can turn **devices** on and off, **starting, operating and closing programmes** and using simple **functions**. (P0)
- Learners can **log in** to a local network or learning environment with their own login. (P0)
- Learners can **store** and **retrieve documents** independently. (P0)
- Learners can handle basic elements of the **user interface** (window, menu, multiple open programmes). (P0)

Grades 3-4, age 8-9:

- Learners can **distinguish** between **operating systems** and **application software**. (P0, P1)
- Pupils know different types of **memory** (e.g., hard disks, flash memory, main memory) and their advantages and disadvantages and understand **size units** for data. (P0, P2)
- Learners can apply **strategies for solving problems** with devices and programmes (e.g., help function, research). (P0, P3)
- Learners can explain how **data can be lost** and know the most important **measures** to protect against this. (P0)

Grades 4-6, age 9-11:

- Learners understand the basic functioning of **search engines**. (P0)
- Learners can distinguish between **local** devices, local **network** and the Internet as **storage locations** for private and public data. (P0, P2)
- Learners have an idea of the **performance units** of information processing systems and can assess their relevance for specific applications (e.g., storage capacity, image resolution, computing capacity, data transmission rate). (P0, P2)

## (C5) Computer Science and Society

*CAS*
Key Stage 2 (grades 3-6, age 7-11)

- **Personal information** should be accurate, stored **securely**, used for **limited purposes** and treated with respect. (CAS, p. 16, Data, Key Stage 2)
    - P2, P3
- Social and **ethical issues** raised by the role of computers in our lives. (CAS, p. 18, Computers, Key Stage 2)
    - P3, P2, possibly P4 & P5

*CSTA*
Level 1: (pre-school – grade 3, age 5-8)

- Practice responsible digital citizenship (legal and **ethical** behaviours) in the use of technology systems and software. (CSTA p.15, L1:3.CI 1.)
- Identify **positive** and **negative** social and **ethical** behaviours for using technology. (CSTA p.15, L1:3.CI 2.)
    - P3, P2, possibly P4
- **Identify jobs** that use computing and technology. (CSTA p. 14, L1:3. CPP 5.)

Level 2: (grades 3-6, age 8-11)

- Understand the **connections** between computer science and **other fields**. (CSTA p. 13, L1:6. CT 6.)
    - P2, P3
- Identify a wide range of jobs that require knowledge or use of computing. (CSTA p. 14, L1:6. CPP 9.)
    - P3

- Understand the **pervasiveness** of computers and computing in daily life (e.g., voice mail, downloading videos and audio files, microwave ovens, thermostats, wireless Internet, mobile computing devices, GPS systems). (CSTA p.14, L1: 6. CT 2.)
    - P3
- Identify factors that **distinguish humans from machines**. (CSTA p.14, L1: 6. CT 5.)
    - P3, possibly P2
- Recognise that computers **model intelligent behaviour** (as found in robotics, speech and language recognition, and computer animation). (CSTA p.15, L1:6. CT 6.)
    - P3, possibly P2
- Discuss basic issues related to **responsible use** of technology and information, and the consequences of inappropriate use. (CSTA p.15, L1:6. CI 1.)
    - P3, P2, possibly P4
- Identify the **impact** of technology (e.g., social networking, cyber bullying, mobile computing and communication, web technologies, cyber security, and virtualisation) on **personal life and society**. (CSTA p.15, L1:6.CI 2.)
    - P3, P2, possibly P4
- **Evaluate** the accuracy, relevance, appropriateness, comprehensiveness, and biases that occur in electronic information sources. (CSTA p.15, L1:6.CI 3.)
    - P2
- Understand ethical issues that relate to computers and networks (e.g., equity of access, security, privacy, copyright, and intellectual property). (CSTA p.15, L1:6.CI 4.)
    - P2, possibly P3

*New Zealand*
Level 1 (grades 1-3, age 5-7)

- **Ergonomics,** digital devices in everyday life; **ethical** and **safe use** (P0)
- Use technology safely and respectfully, **keeping** personal information **private** (P0)
- Identify **where to go for help** and support when they have concerns about content or contact on the internet or other online technologies (P0, P4?)
- **Recognise** use beyond school (P0, P3?)

Level 2 (grades 3-5, age 7-9)

- **Use** safely, respectfully and responsibly; appropriate behaviour; reporting concerns (P0, P2?)
- How systems meet community and personal needs; **evaluate adequacy** of a solution (P0, P2, P3)
- **Collaboratively plan creation** and communication of information (P4)
- **Ethical decisions** and behaviour; social media (P2, P3)

*Swiss curriculum 21*
Learners understand the **structure** and **functioning** of concepts of information processing and apply concepts of **secure data processing**.

Grades 4-6, age 9-11:

- Learners understand the basic functioning of **search engines**.
- Learners can distinguish between **local** devices, local **network** and the Internet as **storage locations** for private and public data.

## Illustration Credits

Seite 11 © Bettina Volke/"Haus der kleinen Forscher" Foundation
Seite Titelbild, 19, 33, 37, 41, 61, 65, 85, 99, 106, 140, 146, 173, 179, 193, 227, 234, 237, 269, 287, 294, 297, 319 © Christoph Wehrer/"Haus der kleinen Forscher" Foundation
Seite 54, 76, 127, 258 © "Haus der kleinen Forscher" Foundation
Seite 241, 253, 256, 260, 262 © InfoSphere-Team
Seite 244, 246, 249, 252 © Kathrin Müller
Seite 282 Illustration: Tim Brackmann, Berlin / © "Haus der kleinen Forscher" Foundation

## "Haus der kleinen Forscher" Foundation

The non-profit "Haus der kleinen Forscher" Foundation (Little Scientists' House) is Germany's largest early childhood education initiative in the domains of science, technology, engineering/computer science, and mathematics (STEM). With an accompanying focus on Education for Sustainable Development (ESD), the aim of the programme is to strengthen children for the future, provide them with important skills, and enable them to act in a sustainable way. Together with its local network partners, the Foundation provides a nationwide continuing professional development programme that supports pedagogical staff at early childhood education and care centres, after-school centres, and primary schools in facilitating the exploration, inquiry, and learning of children between the ages of three and ten. To this end, the "Haus der kleinen Forscher" Foundation improves educational opportunities, fosters interest in the domains of science, technology, computer science, and mathematics, and professionalises pedagogical staff. The partners of the "Haus der kleinen Forscher" Foundation are the Siemens Stiftung, the Dietmar Hopp Stiftung, the Dieter Schwarz Foundation, and the Friede Springer Stiftung. The Foundation is supported by the German Federal Ministry of Education and Research (BMBF).

> ### Vision and Mission of the "Haus der kleinen Forscher" Foundation
>
> **Our Vision: Questioning – Inquiring – Shaping the Future**
>
> Our vision is that all children in Germany will experience educational venues where they can pursue their own questions and explore the world around them in an inquiry-based way. These "Little Scientists' Houses" will strengthen children for the future and empower them to think for themselves and to act responsibly.
>
> Technologisation, digitisation, and the consequences of climate change and social inequality increasingly influence our everyday lives. We shall contribute to enabling people to find their bearings in our rapidly changing world and to remain open to new things.
>
> Everyday engagement with nature and technology fosters children's enjoyment of learning and thinking. We see early education as a key to being able to successfully meet the challenges of a complex world.

**Our Mission**

The mission of the "Haus der kleinen Forscher" Foundation is to …

- promote a questioning and inquiring attitude in children;
- give children the opportunity to discover at a young age their own talents and potential in the domains of science, technology, computer science, and mathematics; and
- lay the foundations for reflective engagement with technological and social changes in the interests of sustainable development.

Together with their persons of reference, the children experience fun and enjoyment in exploring and understanding the world around them. Children actively shape their education processes, thereby experiencing themselves as competent and self-efficacious in their everyday lives. In the course of inquiry-based learning, children can develop problem-solving skills, find their own answers, and gain a feeling of self-confidence ("Yes, I can!"). The importance of these experiences and abilities for personality development and the child's future professional biography extends far beyond childhood.

With a practice-oriented and high-quality approach to professionalisation, the Foundation supports early childhood professionals and primary teachers in facilitating the exploration, inquiry, and learning activities of children up to the age of ten. Through diverse continuing professional development offerings, early childhood professionals and primary teachers experience for themselves the fascination of engaging in independent inquiry. They expand their knowledge and pedagogical competencies, and implement them in their everyday work with the children.

The initiative supports educational institutions in sustainably developing themselves as "venues of inquiry-based learning" and – as "Little Scientists' Houses" – in creating favourable learning environments for children.

# English Publications issued by the "Haus der kleinen Forscher" Foundation to date

## Early Science Education – Goals and Process-Related Quality Criteria for Science Teaching (2018)

*Yvonne Anders, Ilonca Hardy, Sabina Pauen, Jörg Ramseger, Beate Sodian, Mirjam Steffensky, Russell Tyler*

The fifth volume in the series "Scientific Studies on the Work of the 'Haus der kleinen Forscher' Foundation" focuses on goals of science education at the level of the children, the early childhood professionals, and the pedagogical staff at after-school centres and primary schools, and on process-related quality criteria for science teaching at the pre-primary and primary level.

In their expert reports, Yvonne Anders, Ilonca Hardy, Sabina Pauen, Beate Sodian, and Mirjam Steffensky specify pedagogical content dimensions of the goals of early science education at pre-primary and primary school age. In addition to theoretically underpinning these goals, the authors present instruments for their assessment. In his expert report, Jörg Ramseger formulates ten quality criteria for science teaching. Early childhood professionals and pedagogical staff at after-school centres and primary schools can draw on these process-related criteria when planning lessons and conducting self-evaluations of science learning opportunities at the pre-primary and primary level. The concluding chapter of the volume describes the implementation of these expert recommendations in the substantive offerings of, and the accompanying research on, the "Haus der kleinen Forscher" Foundation.

## Using Science to Do Social Good: STEM Education for Sustainable Development (2019)

*Janna Pahnke, Carol O'Donnell, & MartínBascopé, M.*

Position paper developed in preparation for the second "International Dialogue on STEM Education" (IDoS) in Berlin, December 5-6, 2019.

The international position paper argues for an integrated approach of STEM Education for Sustainable Development. It analyses critically how an integrated and transdisciplinary focus on inquiry-based STEM education could serve to enhance sustainable development and build capacity for future generations. As such, the international paper promotes the idea of a transdisciplinary framework of education, acknowledging the complex context of global challenges and the need for integrating values, ethics, and world views towards the development of sustainable mindsets and using science to do social good. After reviewing the context and pedagogical basis of this approach, the paper presents a set of goals and guiding principles of STEM Education for Sustainable Development.

Available here (in English and Spanish): https://www.haus-der-kleinen-forscher.de/en/international-dialogue-on-stem-education/publications